Frontiers in Mathematics

Advisory Editorial Board

Alexandru Aleman
Nathan S. Feldman
William T. Ross

The
Hardy Space
of a
Slit
Domain

Birkhäuser
Basel · Boston · Berlin

Author:

Alexandru Aleman
Centre for Mathematical Sciences
Lund University
221 00 Lund
Sweden
e-mail: Alexandru.Aleman@math.lu.se

Nathan S. Feldman
Department of Mathematics
Washington & Lee University
Lexington, VA 24450
USA
e-mail: feldmanN@wlu.edu

William T. Ross
Department of Mathematics and
Computer Science
University of Richmond
Richmond, VA 23173
USA
e-mail: wross@richmond.edu

2000 Mathematical Subject Classification: 30D55, 47A15, 47A16

Library of Congress Control Number: 2009931266

Bibliographic information published by Die Deutsche Bibliothek
Die Deutsche Bibliothek lists this publication in the Deutsche Nationalbibliografie;
detailed bibliographic data is available in the Internet at <http://dnb.ddb.de>.

ISBN 978-3-0346-0097-2 Birkhäuser Verlag AG, Basel · Boston · Berlin

© 2009 Birkhäuser Verlag AG
Basel · Boston · Berlin
P.O. Box 133, CH-4010 Basel, Switzerland
Part of Springer Science+Business Media
Cover design: Birgit Blohmann, Zürich, Switzerland
Printed on acid-free paper produced from chlorine-free pulp. TCF ∞
Printed in Germany

ISBN 978-3-0346-0097-2   e-ISBN 978-3-0346-0098-9

9 8 7 6 5 4 3 2 1   www.birkhauser.ch

Alexandru Aleman
Nathan S. Feldman
William T. Ross

The
Hardy Space
of a
Slit
Domain

Birkhäuser
Basel · Boston · Berlin

Author:

Alexandru Aleman
Centre for Mathematical Sciences
Lund University
221 00 Lund
Sweden
e-mail: Alexandru.Aleman@math.lu.se

Nathan S. Feldman
Department of Mathematics
Washington & Lee University
Lexington, VA 24450
USA
e-mail: feldmanN@wlu.edu

William T. Ross
Department of Mathematics and
Computer Science
University of Richmond
Richmond, VA 23173
USA
e-mail: wross@richmond.edu

2000 Mathematical Subject Classification: 30D55, 47A15, 47A16

Library of Congress Control Number: 2009931266

Bibliographic information published by Die Deutsche Bibliothek
Die Deutsche Bibliothek lists this publication in the Deutsche Nationalbibliografie;
detailed bibliographic data is available in the Internet at <http://dnb.ddb.de>.

ISBN 978-3-0346-0097-2 Birkhäuser Verlag AG, Basel · Boston · Berlin

© 2009 Birkhäuser Verlag AG
Basel · Boston · Berlin
P.O. Box 133, CH-4010 Basel, Switzerland
Part of Springer Science+Business Media
Cover design: Birgit Blohmann, Zürich, Switzerland
Printed on acid-free paper produced from chlorine-free pulp. TCF ∞
Printed in Germany

ISBN 978-3-0346-0097-2   e-ISBN 978-3-0346-0098-9

9 8 7 6 5 4 3 2 1   www.birkhauser.ch

Preface

If $\mathcal{H}$ is a Hilbert space and $T : \mathcal{H} \to \mathcal{H}$ is a continous linear operator, a natural question to ask is: What are the closed subspaces $\mathcal{M}$ of $\mathcal{H}$ for which $T\mathcal{M} \subset \mathcal{M}$? Of course the famous invariant subspace problem asks whether or not T has *any* non-trivial invariant subspaces. This monograph is part of a long line of study of the invariant subspaces of the operator $T = M_z$ (multiplication by the independent variable z, i.e., $M_z f = zf$) on a Hilbert space of analytic functions on a bounded domain G in $\mathbb{C}$. The characterization of these M_z-invariant subspaces is particularly interesting since it entails both the properties of the functions inside the domain G, their zero sets for example, as well as the behavior of the functions near the boundary of G. The operator M_z is not only interesting in its own right but often serves as a model operator for certain classes of linear operators. By this we mean that given an operator T on $\mathcal{H}$ with certain properties (certain subnormal operators or two-isometric operators with the right spectral properties, etc.), there is a Hilbert space of analytic functions on a domain G for which T is unitarity equivalent to M_z.

Probably the first to successfully study these types of problems was Beurling [13] who gave a complete characterization of the M_z-invariant subspaces of the Hardy space of the unit disk. These are the functions $f(z) = \sum_{n=0}^{\infty} a_n z^n$ which are analytic on the open unit disk $\mathbb{D} := \{|z| < 1\}$ for which $\sum_{n \geqslant 0} |a_n|^2 < \infty$. Many others followed with a discussion, often a complete characterization, of the M_z-invariant subspaces where the Hardy space is replaced by the space of analytic functions $f(z) = \sum_{n=0}^{\infty} a_n z^n$ on $\mathbb{D}$ satisfying $\sum_{n \geqslant 0} w_n |a_n|^2 < \infty$, where $(w_n)_{n \geqslant 0}$ is a sequence of positive weights. For example, when $w_n = n$, we get the classical Dirichlet space where the M_z-invariant subspaces were discussed in [60, 61, 62]. When $w_n = n^\alpha$ and $\alpha > 1$, we get certain weighted Dirichlet spaces where the M_z-invariant subspaces were completely characterized in [69]. See [52, 53] for some related results. When $w_n = n^{-1}$ (or more generally $w_n = n^\alpha, \alpha < 0$), we get the Bergman (weighted Bergman) spaces where the M_z-invariant subspaces were discussed in [8, 68]. See also [30, 42].

In Beurling's seminal paper, and the ones that followed, notice how the underlying domain of analyticity is kept fixed to be the unit disk $\mathbb{D}$, but the Hilbert space of analytic functions is changed by varying the weights w_n. In a series of papers beginning with Sarason [65], the basic type of Hilbert space is fixed but the domain of analyticity is changed. To see what we mean here, the condition $f(z) = \sum_{n \geqslant 0} a_n z^n$ is analytic on $\mathbb{D}$ and $\sum_{n \geqslant 0} |a_n|^2 < \infty$, the definition of the Hardy space of $\mathbb{D}$, can be equivalently restated as

f is analytic on $\mathbb{D}$ and there is a harmonic function U on $\mathbb{D}$ for which $|f|^2 \leqslant U$ on $\mathbb{D}$. Such a function U is called a harmonic majorant for $|f|^2$. For a general bounded domain $G \subset \mathbb{C}$, one can define the Hardy space of G to be the analytic functions f on G for which $|f|^2$ has a harmonic majorant on G. Beginning with Sarason's paper, there were several authors [6, 7, 37, 44, 64, 76, 77, 78] who characterized the M_z-invariant subspaces of the Hardy space of annular-type domains, which include an annulus, a disk with several holes removed, and a crescent domain (the region between two internally tangent circles).

Conspicuously missing from this list of domains are slit domains, for example $G = \mathbb{D} \setminus [0,1)$. In this monograph, we obtain a complete characterization of the M_z-invariant subspaces of the Hardy space of slit domains. Along the way, we give a thorough exposition of the Hardy space, and even the Hardy-Smirnov space, of a slit domain as well as several applications of our results to de Branges-type spaces and the classical backward shift operator of the Hardy space of $\mathbb{D}$. We also discuss several aspects of the operator $M_z|\mathcal{M}$, where $\mathcal{M}$ is an M_z-invariant subspace of the Hardy space of G. In particular, we explore questions about cyclicity, the spectrum, and the essential spectrum for $M_z|\mathcal{M}$.

Contents

Notation

The complete list of symbols is contained in the next chapter. Below are some basic symbols and remarks regarding notation and organization.

- $\mathbb{C}$ (complex numbers)
- $\widehat{\mathbb{C}} = \mathbb{C} \cup \{\infty\}$ (Riemann sphere)
- $\mathbb{R}$ (real numbers)
- $\mathbb{D} = \{z \in \mathbb{C} : |z| < 1\}$
- $\mathbb{T} = \partial \mathbb{D} = \{z \in \mathbb{C} : |z| = 1\}$
- $\mathbb{N} = \{1, 2, \ldots\}$
- $\mathbb{N}_0 = \{0, 1, 2, \ldots\}$
- When defining functions, sets, operators, etc., we will often use the notation $A :=$ *xxx*. By this we mean A 'is defined to be' *xxx*.
- As is traditional in analysis, the constants $c, c', c'', \ldots c_1, c_2, \ldots$ can change from one line to the next without being relabeled.
- Numbering is done by chapter and section, and *all* equations, theorems, propositions, and such are numbered consecutively.
- If J is a set in some topological vector space, $\bigvee J$ is the closed linear span of the elements of J and J^- is the closure of J.
- If $A \subset \mathbb{C}$, then $\overline{A} := \{\overline{a} : a \in \mathbb{C}\}$ is the complex conjugate of the elements of A. From the previous item, note that A^- is the closure of A.
- A *linear manifold* in some topological vector space is a set which is closed under the basic vector space operations. A *subspace* is a closed (topologically) linear manifold.

List of Symbols

Preamble

The statement of our main results, as well as the techniques used to prove them, become more meaningful if we review the basics of the Hardy space of the open unit disk $\mathbb{D}$. The reader familiar with this material can skip this section. Several standard texts are [31, 34, 45, 51, 57].

For $0 < p < \infty$, let H^p, *the Hardy space*, denote the space of functions f analytic on $\mathbb{D}$ for which the L^p integral means

$$M_p(r;f) := \left\{ \int_0^{2\pi} |f(re^{i\theta})|^p \frac{d\theta}{2\pi} \right\}^{1/p}$$

remain bounded as $r \uparrow 1^-$. This definition can be extended to $p = \infty$ by

$$M_\infty(r;f) := \sup\{|f(re^{i\theta})| : \theta \in [0, 2\pi]\}$$

and so H^∞ is the set of bounded analytic functions on $\mathbb{D}$. The function $r \mapsto M_p(r;f)$ is increasing on the interval $[0,1)$ and the quantity

$$\|f\|_{H^p} := \sup_{0<r<1} M_p(r;f) = \lim_{r \uparrow 1^-} M_p(r;f)$$

defines a norm on H^p when $1 \leqslant p < \infty$. When $0 < p < 1$, the quantity

$$\text{dist}(f,g) := \|f - g\|_{H^p}^p$$

defines a translation invariant metric on H^p. The pointwise estimate

$$|f(z)| \leqslant 2^{1/p} \|f\|_{H^p} \frac{1}{(1-|z|)^{1/p}}, \quad z \in \mathbb{D},$$

can be used to show that H^p ($1 \leqslant p < \infty$) is a Banach space while H^p ($0 < p < 1$) is an F-space (a complete translation invariant metric space). In particular, if $f_n \to f$ in H^p, then $f_n \to f$ uniformly on compact subsets of $\mathbb{D}$. For $p = 2$, H^2 becomes a Hilbert space with inner product

$$\langle f,g \rangle = \sum_{n=0}^\infty a_n \overline{b_n},$$

where $(a_n)_{n \geqslant 0}$ are the Taylor coefficients (about the origin) of f and $(b_n)_{n \geqslant 0}$ are those of g. Here is a collection of standard facts about H^p. Proofs of the results below, as well as the rest of the material from this chapter, are found in [31].

Theorem. *For $0 < p \leqslant \infty$ and $f \in H^p$,*

1.

$$f(e^{i\theta}) := \lim_{r \to 1^-} f(re^{i\theta})$$

exists for almost every θ.

2. *The almost everywhere defined boundary function $e^{i\theta} \mapsto f(e^{i\theta})$ belongs to L^p and when $0 < p < \infty$,*

$$\lim_{r \to 1^-} \int_0^{2\pi} |f(re^{i\theta}) - f(e^{i\theta})|^p \frac{d\theta}{2\pi} = 0.$$

Hence $\|f\|_{H^p} = \|f\|_{L^p}$.

3. *If $f \in H^p \setminus \{0\}$, then*

$$\int_0^{2\pi} \log |f(e^{i\theta})| \frac{d\theta}{2\pi} > -\infty$$

and hence the function $e^{i\theta} \mapsto f(e^{i\theta})$ can not vanish on any set of positive measure.

4. *If $p \geqslant 1$, and $f \in H^p$ has Taylor series*

$$f(z) = \sum_{n=0}^{\infty} a_n z^n,$$

then

$$a_n = \int_0^{2\pi} f(e^{i\theta}) e^{-in\theta} \frac{d\theta}{2\pi}, \quad n \in \mathbb{N}_0.$$

5. *If $p \geqslant 1$ and $f \in H^p$, we have the Cauchy integral formula*

$$f(z) = \frac{1}{2\pi i} \oint_{\mathbb{T}} \frac{f(\zeta)}{\zeta - z} d\zeta,$$

where $\mathbb{T} := \partial \mathbb{D}$.

6. *For $0 < p < \infty$, the polynomials are dense in H^p. When $p = \infty$, the polynomials are weak-$*$ dense[1] in H^∞.*

From our collection of facts about H^2, one can show that the inner product on H^2 can be written as

$$\langle f, g \rangle = \int_0^{2\pi} f(e^{i\theta}) \overline{g(e^{i\theta})} \frac{d\theta}{2\pi}.$$

Theorem (Smirnov). *If $0 < p < q$ and $f \in H^p$ has L^q boundary values, then $f \in H^q$.*

---

[1] See [34, p. 85] for more on the weak-$*$ topology on H^∞.

We know that, via boundary functions, H^p can be viewed as a closed subspace of L^p. Turning this problem around, one can ask: when does a given $f \in L^p$ belong to H^p? At least for $p \geqslant 1$, there is an answer given by a theorem of F. and M. Riesz.

Theorem. *For $p \geqslant 1$, a function $f \in L^p$ belongs to H^p if and only if the Fourier coefficients*

$$\int_0^{2\pi} f(e^{i\theta}) e^{-in\theta} \frac{d\theta}{2\pi}$$

vanish for all $n < 0$.

This result generalized to measures.

Theorem (F. and M. Riesz theorem). *Suppose a finite complex Borel measure μ on $\mathbb{T}$ satisfies*

$$\int_0^{2\pi} e^{in\theta} d\mu(e^{i\theta}) = 0 \quad \text{for all } n \in \mathbb{N}_0.$$

Then $d\mu = \phi \frac{d\theta}{2\pi}$ where $\phi \in H_0^1 = \{f \in H^1 : f(0) = 0\}$.

Every $f \in H^p$ can be factored as

$$f = O_f I_f.$$

The function O_f, the *outer factor*, is characterized by the property that O_f belongs to H^p and

$$\log|O_f(0)| = \int_0^{2\pi} \log|O_f(e^{i\theta})| \frac{d\theta}{2\pi}.$$

Every H^p outer function F (i.e., F has no inner factor) can be expressed as

$$F(z) = e^{i\gamma} \exp\left(\int_0^{2\pi} \frac{e^{i\theta}+z}{e^{i\theta}-z} \log\psi(e^{i\theta}) \frac{d\theta}{2\pi}\right),$$

where γ is a real number, $\psi \geqslant 0$, $\log\psi \in L^1$, and $\psi \in L^p$. Note that F has no zeros in the open unit disk and $|F(e^{i\theta})| = \psi(e^{i\theta})$ almost everywhere. Moreover, every such F as above belongs to H^p and is outer. The *inner factor*, I_f, is characterized by the property that I_f is a bounded analytic function on $\mathbb{D}$ whose boundary values satisfy $|I_f(e^{i\theta})| = 1$ for almost every θ. Furthermore, the inner factor I_f can be factored further as the product of two inner functions

$$I_f = b s_\mu,$$

where b is a *Blaschke product*

$$b(z) = z^m \prod_{n=1}^{\infty} \frac{|a_n|}{a_n} \frac{a_n - z}{1 - \overline{a_n}z}$$

whose zeros at $z = 0$ as well as $\{a_n\} \subset \mathbb{D}\setminus\{0\}$ (repeated according to multiplicity) satisfy the *Blaschke condition*

$$\sum_{n=1}^{\infty} (1 - |a_n|) < \infty,$$

(which guarantees the convergence of the product), and s_μ is the (zero free) *singular inner factor*

$$s_\mu(z) = \exp\left(-\int_0^{2\pi} \frac{e^{i\theta}+z}{e^{i\theta}-z} d\mu(e^{i\theta})\right),$$

where μ is a positive measure on $\mathbb{T}$ which is singular with respect to Lebesgue measure.

A meromorphic function f on $\mathbb{D}$ is said to be of *bounded type* if $f = h_1/h_2$, where h_1, h_2 are bounded analytic functions on $\mathbb{D}$. From our above discussion, a function of bounded type must have finite non-tangential limits almost everywhere on $\mathbb{T}$ and can be factored as

$$f = \frac{I_{h_1} O_{h_1}}{I_{h_2} O_{h_2}}.$$

The set N, the *Nevanlinna class*, will be the functions f of bounded type which are analytic on $\mathbb{D}$ (equivalently I_{h_2} is a singular inner function). The set N^+, the *Smirnov class*, will be the set of $f \in N$ for which I_{h_2} is a constant.

This extension of Smirnov's theorem (see above) due to Polubarinova and Kochina [57, p. 80] (see also [31, p. 28]), will be used many times in this book.

Theorem. *If $f \in N^+$ with L^p boundary function, then $f \in H^p$.*

It turns out that functions in H^p not only have a radial limit almost everywhere on $\mathbb{T}$ but also have a stronger non-tangential limit almost everywhere. We collect some facts about non-tangential limits which will be used at various times in the text.

For $\zeta \in \mathbb{T}$ and $\alpha > 1$, let

$$\Gamma_\alpha(\zeta) := \{z \in \mathbb{D} : |z - \zeta| < \alpha(1 - |z|)\}$$

be a *non-tangential approach region* (often called a *Stoltz region*). Note that $\Gamma_\alpha(\zeta)$ is a triangular shaped region with its vertex at ζ (see Fig. 1).

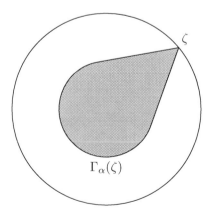

Figure 1: Non-tangential approach region with vertex at $\zeta \in \mathbb{T}$.

We say that f has a *non-tangential limit* value A at ζ, written

$$\angle \lim_{z \to \zeta} f(z) = A,$$

if $f(z) \to A$ as $z \to \zeta$ within any non-tangential approach region $\Gamma_\alpha(\zeta)$.

Theorem. *If $f \in H^p$, $0 < p \leqslant \infty$, then f has a finite non-tangential limit almost everywhere on $\mathbb{T}$.*

From our collection of facts about H^p functions, we know that functions in $H^p \setminus \{0\}$ can not have non-tangential limit equal to zero on an any set of positive measure. This fact is not unique to H^p functions.

Theorem (Privalov's uniqueness theorem [17, 51, 57]). *Suppose f is analytic on $\mathbb{D}$ and*

$$\angle \lim_{z \to \zeta} f(z) = 0$$

for ζ in some subset of $\mathbb{T}$ of positive Lebesgue measure. Then $f \equiv 0$.

Non-tangential limits are important in the statement of Privalov's theorem since there are non-trivial analytic functions on $\mathbb{D}$ which have radial limits equal to zero almost everywhere on $\mathbb{T}$ [12].

Since invariant subspaces is the heart of this book, let us mention Beurling's theorem. The shift operator $S : H^2 \mapsto H^2$, defined by

$$(Sf)(z) = zf(z),$$

is an isometry on H^2. A classical theorem of A. Beurling [13] (see also [31]) characterizes the invariant subspaces of S. By 'invariant subspace' we mean a closed linear manifold $\mathcal{M} \subset H^2$ for which $S\mathcal{M} \subset \mathcal{M}$. If ϑ is an inner function, then $\|\vartheta f\| = \|f\|$ for all $f \in H^2$ and so ϑH^2 is a closed linear manifold (a subspace) of H^2. It is also clearly S-invariant. Beurling's theorem says these are all of them.

Theorem (Beurling). *If ϑ is an inner function, the set ϑH^2 is an S-invariant subspace of H^2. Conversely, if $\mathcal{M} \subset H^2$, $\mathcal{M} \neq \{0\}$, is an S-invariant subspace, then $\mathcal{M} = \vartheta H^2$ for some inner function ϑ.*

Since the main purpose of this book is to essentially prove a version of Beurling's theorem for the Hardy space of a slit disk, we include a proof of Beurling's theorem for the disk.

Proof of Beurling's theorem. The proof that ϑH^2 is an S-invariant subspace of H^2 was discussed in our preliminary remarks. To prove the second part of the theorem, suppose $\mathcal{M}$ is a non-zero S-invariant subspace of H^2. First notice that $S\mathcal{M} \neq \mathcal{M}$. If this were not the case, then $f/z \in \mathcal{M}$ whenever $f \in \mathcal{M}$. Applying this k times we conclude that

$$\frac{f}{z^k} \in \mathcal{M} \quad \forall k \in \mathbb{N}.$$

But this would mean, since f/z^k must be analytic on $\mathbb{D}$, that $f \equiv 0$, a contradiction to the assumption that $\mathcal{M} \neq \{0\}$.

Second, since S is an isometry, $S\mathcal{M}$ is closed and since $S\mathcal{M} \neq \mathcal{M}$, one observes that

$$\mathcal{M} \cap (S\mathcal{M})^{\perp} \neq \{0\}.$$

Thus $\mathcal{M} \cap (S\mathcal{M})^{\perp}$ contains a non-trivial function ϑ. We now argue that $|\vartheta| = c$ on a set of full measure in $\mathbb{T}$. Indeed,

$$\int_0^{2\pi} |\vartheta(e^{i\theta})|^2 e^{-in\theta} \frac{d\theta}{2\pi} = \langle \vartheta, S^n \vartheta \rangle = 0 \quad \forall n \in \mathbb{N}.$$

Taking complex conjugates of both sides of the above equation, we also see that

$$\int_{\mathbb{T}} |\vartheta(e^{i\theta})|^2 e^{in\theta} \frac{d\theta}{2\pi} = 0 \quad \forall n \in \mathbb{N}.$$

This means that the Fourier coefficients of $|\vartheta|^2$ all vanish except for $n = 0$ and so $|\vartheta|^2 = c$ almost everywhere on $\mathbb{T}$. Without loss of generality, we can assume that $|\vartheta| = 1$ almost everywhere on $\mathbb{T}$ and so ϑ is an inner function.

Third, let $[\vartheta]$ denote the closed linear span of the functions

$$\vartheta, S\vartheta, S^2\vartheta, \ldots$$

and observe that

$$[\vartheta] = \vartheta H^2.$$

To see this, notice that clearly $[\vartheta] \subset \vartheta H^2$. For the other containment, let $g = \vartheta G \in \vartheta H^2$ and let G_N be the N-th partial sum of the Taylor series of G. Notice that $\vartheta G_N \in [\vartheta]$ since G_N is a polynomial. From Parseval's theorem, $G_N \to G$ in H^2 and so, since ϑ is a bounded function, ϑG_N converges to ϑG in H^2.

Finally, observe that

$$[\vartheta] = \mathcal{M}.$$

Indeed, $\vartheta \in \mathcal{M}$ and so $[\vartheta] \subset \mathcal{M}$. Now suppose that $f \in \mathcal{M}$ and $f \perp [\vartheta]$. Since $f \perp [\vartheta]$,

$$\int_0^{2\pi} f(e^{i\theta}) \overline{\vartheta(e^{i\theta})e^{in\theta}} \frac{d\theta}{2\pi} = \langle f, S^n \vartheta \rangle = 0 \quad \forall n \in \mathbb{N}_0.$$

But since $\vartheta \perp S\mathcal{M}$, we also know that

$$\int_0^{2\pi} f(e^{i\theta}) \overline{\vartheta(e^{i\theta})} e^{in\theta} \frac{d\theta}{2\pi} = \langle S^n f, \vartheta \rangle = 0 \quad \forall n \in \mathbb{N}.$$

The previous two equations say that all of the Fourier coefficients of $f\overline{\vartheta}$ vanish and so $f\overline{\vartheta} = 0$ almost everywhere on $\mathbb{T}$. But we have already shown that $|\vartheta| = 1$ almost everywhere on $\mathbb{T}$ and so $f \equiv 0$. $\qquad \square$

The key to proving Beurling's theorem is the fact that the invariant subspace generated by $\mathcal{M} \cap (S\mathcal{M})^{\perp}$ is equal to $\mathcal{M}$. This idea extends to other Hilbert spaces of analytic functions [8, 60, 68], but not to the Hardy space of a slit domain. There is a Beurling theorem for the H^p spaces [31, 34]: suppose $0 < p < \infty$ and $\mathcal{M}$ is a non-zero subspace of H^p. Then $\mathcal{M}$ is S-invariant if and only if $\mathcal{M} = \vartheta H^p$ for some inner function ϑ.

Chapter 1

Introduction

1.1 Some history

This monograph continues the study of the *invariant subspaces* of the *Hardy space $H^2(\Omega)$* of a bounded domain $\Omega \subset \mathbb{C}$ (see Chapter 2 for a definition of the Hardy space). By the term 'invariant subspace', we mean a subspace (i.e., a closed linear manifold) $\mathcal{M}$ of $H^2(\Omega)$ for which $S\mathcal{M} \subset \mathcal{M}$, where $(Sf)(z) = zf(z)$ is the operator 'multiplication by z'.

When Ω is the open unit disk $\mathbb{D} := \{z \in \mathbb{C} : |z| < 1\}$, a much celebrated theorem of Beurling ([22, p. 135] or [34, p. 82]) says that every non-trivial invariant subspace $\mathcal{M}$ takes the form $\mathcal{M} = \Theta H^2(\mathbb{D})$, where Θ is an *inner function* meaning that Θ is a bounded analytic function on $\mathbb{D}$ whose non-tangential boundary function has constant modulus one almost everywhere on $\partial \mathbb{D}$. A similar result is true when Ω is a Jordan domain with smooth boundary [32]. When Ω is a finitely connected (bounded) domain with disjoint analytic boundary contours – an annulus for example – the invariant subspaces of $H^2(\Omega)$ were examined in a series of papers beginning with Sarason [65] and continuing with Hasumi [37], Voichick [76, 77], Royden [64], Hitt [44], Yakubovich [78], Aleman and Richter [7], Aleman and Olin [6]. Due to the 'holes' in the region, there are several types of invariant subspaces to consider. First there are the 'fully invariant' subspaces $\mathcal{M}$ which, by definition, satisfy $r\mathcal{M} \subset \mathcal{M}$ for every rational function r whose poles are off of Ω^-. In this case [37, 65, 76, 77], $\mathcal{M} = \Theta H^2(\Omega)$, where Θ is an inner function on Ω meaning that Θ is a bounded analytic function on Ω whose non-tangential boundary function has constant modulus on each of the components of $\partial \Omega$. Then there are the invariant subspaces $\mathcal{M}$ for which $r\mathcal{M} \subset \mathcal{M}$ for rational functions r whose poles lie in certain components of $\mathbb{C} \setminus \Omega^-$ but not others. In this case [7, 44, 64, 78], the description of these $\mathcal{M}$ depends on, amongst other things, the behavior of the pseudocontinuation of functions across the holes.

The purpose of this monograph is to broaden the discussion to include (simply connected) slit domains,

$$\Omega := \mathbb{D} \setminus \bigcup_{j=1}^{N} \gamma_j, \qquad (1.1.1)$$

where $\gamma_1, \ldots, \gamma_N$ are simple disjoint analytic arcs (see Chapter 10 for the precise technical restrictions on the arcs). See Figure 1.1 for an example.

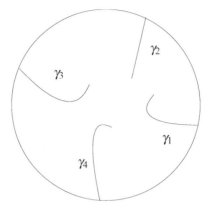

Figure 1.1: A (simply connected) slit domain Ω.

For example, perhaps the simplest type of slit domain to consider here is $G = \mathbb{D} \setminus [0,1]$ (see Figure 1.2).

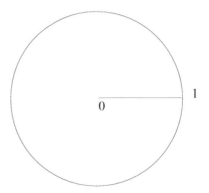

Figure 1.2: The slit domain $G = \mathbb{D} \setminus [0,1)$.

1.2   Invariant subspaces of the slit disk

Before stating one of our main theorems, the complete description of the invariant subspaces of $H^2(\Omega)$ for slit domains, let us give some examples. For clearer exposition, we state these examples, as well as our main results, for the simple slit domain G. The results for the more general slit domains Ω in (1.1.1) are stated in Chapter 10.

The description of the invariant subspaces of $H^2(G)$ depends on, amongst other things, the behavior of the functions near the boundary of G. Here is where we encounter our first major difference between the case discussed previously, where the domain was an annular type domain, and our current case of the slit domain G. In the former case, when computing the boundary function, the boundary is accessible from only one direction. In the latter case however, part of the boundary – the slit – is accessible from two directions – from the top and from the bottom of the slit. For each $f \in H^2(G)$, the functions

$$f^+(x) := \lim_{y \to 0^+} f(x+iy) \quad \text{and} \quad f^-(x) := \lim_{y \to 0^-} f(x+iy)$$

are defined for almost every $x \in [0,1]$ and $f^+, f^- \in L^2([0,1], \omega)$, where ω is the harmonic measure for G. Furthermore,

$$f(e^{i\theta}) := \lim_{r \to 1^-} f(re^{i\theta})$$

exists for almost every θ and this function belongs to $L^2(\mathbb{T}, \omega)$, where $\mathbb{T} = \partial \mathbb{D}$. In fact (Proposition 2.3.4), the norm on $H^2(G)$ satisfies

$$\|f\|_{H^2(G)}^2 = \int_{[0,1]} \left(|f^+|^2 + |f^-|^2\right) d\omega + \int_{\mathbb{T}} |f|^2 d\omega.$$

We will discuss these preliminaries in Chapter 2.

The first type of invariant subspaces $\mathcal{M} \subset H^2(G)$ to consider are those for which $H^\infty(G)\mathcal{M} \subset \mathcal{M}$, where $H^\infty(G)$ is the set of bounded analytic functions on G. By realizing that

$$H^2(G) = \left\{f \circ \phi_G^{-1} : f \in H^2(\mathbb{D})\right\},$$

where ϕ_G is a conformal map from $\mathbb{D}$ onto G (see the appendix for a specific formula), and using Beurling's theorem, we can characterize these $H^\infty(G)$-invariant subspaces by means of 'inner' functions. We say a bounded analytic function Θ on G is G-*inner* if $\Theta \circ \phi_G$ is inner in the classical sense (the almost everywhere defined boundary function on $\partial \mathbb{D}$ has constant modulus 1 almost everywhere), or equivalently, the boundary function defined by Θ^+, Θ^-, and $\Theta|\mathbb{T}$, as above, has modulus 1 almost everywhere. These $H^\infty(G)$-invariant subspaces are characterized in Proposition 6.1.2 as follows.

Theorem 1.2.1. *Suppose $\mathcal{M}$ is a non-trivial subspace of $H^2(G)$ such that*

$$H^\infty(G)\mathcal{M} \subset \mathcal{M}. \tag{1.2.2}$$

Then there is a G-inner function Θ such that $\mathcal{M} = \Theta H^2(G)$. Moreover, for every G-inner function Θ, the subspace $\mathcal{M} := \Theta H^2(G)$ satisfies (1.2.2).

The subspaces $\Theta H^2(G)$, where Θ is G-inner, are not all of the invariant subspaces of $H^2(G)$. To see this, we need only consider the class of invariant subspaces

$$\mathcal{M}(\rho, E) := \left\{f \in H^2(G) : f^+ = \rho f^- \text{ a.e. on } E\right\},$$

where $\rho : [0,1] \to \mathbb{C}$ is a measurable function and E is a measurable subset of $[0,1]$. Since convergence of a sequence in the $H^2(G)$-norm implies a subsequence converges almost everywhere on ∂G, we see that $\mathcal{M}(\rho, E)$ is closed in $H^2(G)$ and, since the identity function $z \mapsto z$ is analytic across the slit $[0, 1]$, $\mathcal{M}(\rho, E)$ is clearly invariant. The subspace $\mathcal{M}(\rho, E)$ is never equal to $\Theta H^2(G)$ for some G-inner function Θ since $\mathcal{M}(\rho, E)$ can not contain both Θ and $\Theta\sqrt{z}$.

Perhaps every invariant subspace takes the form $\Theta\mathcal{M}(\rho, E)$. Though this seems reasonable, it is not the case. Consider the functions from $H^2(G)$ which have an analytic continuation across $[0, 1)$. We will show in Corollary 6.1.8 that this class of functions is closed in $H^2(G)$ and is equal to $\mathcal{M}(1, [0, 1])$. In fact, see (7.3.1) and Theorem 7.3.2, the unit ball in this space forms a normal family on $\mathbb{D}$. Hence

$$\mathcal{M}_0 := \{f \in \mathcal{M}(1, [0, 1]) : f(0) = 0\}$$

is closed in $H^2(G)$ and is an invariant subspace of $H^2(G)$. Moreover, since the common zero of $\mathcal{M}_0$ is at the origin, which is on the slit and not in G (where one could take it into account using the G-inner function Θ), this space is not of the form $\Theta\mathcal{M}(\rho, E)$. Note that $\mathcal{M}_0$ can be equivalently described as

$$\left\{f \in \mathcal{M}(1, [0, 1]) : \frac{f}{z} \in H^2(G)\right\}.$$

Furthermore, the function $F(z) = z$ is G-*outer* in the sense that $F \circ \phi_G$ is outer in the classical sense of

$$\log|F \circ \phi_G(0)| = \int_{\mathbb{T}} \log|F \circ \phi_G(\zeta)|\frac{|d\zeta|}{2\pi}.$$

These examples indicate that the description of the invariant subspaces depends on the *four* parameters Θ, ρ, E, F. Our main theorem (Corollary 6.1.6 and Theorem 6.2.1) codifies this as follows: For $\varepsilon \in (0, 1)$, let $G_\varepsilon := \mathbb{D} \setminus [-\varepsilon, 1]$.

Theorem 1.2.3. *Let $\mathcal{M}$ be a non-trivial invariant subspace of $H^2(G)$ with greatest common G-inner divisor $\Theta_\mathcal{M}$. Then there exists a measurable set $E \subset [0, 1]$, a measurable function $\rho : [0, 1] \to \mathbb{C}$ and, given any $\varepsilon \in (0, 1)$, there exists a G_ε-outer function F_ε such that*

$$\mathcal{M} = \Theta_\mathcal{M} \cdot \left\{f \in H^2(G) : \frac{f}{F_\varepsilon} \in H^2(G_\varepsilon), f^+ = \rho f^- \text{ a.e. on } E\right\}. \tag{1.2.4}$$

An alert reader might wonder why such a linear manifold defined by the right-hand side of (1.2.4) is actually closed, i.e., a subspace. Theorem 1.2.3 does not say that for *any* G_ε-outer functions F_ε the linear manifold in (1.2.4) is closed. It just says that given $\mathcal{M}$ there are *some* G_ε-outer functions F_ε for which $\mathcal{M}$ can be written in terms of these particular F_ε's (and the other parameters $\Theta_\mathcal{M}, \rho$, and E). Furthermore, the F_ε's are not unique. The set in (1.2.4) remains unchanged if F_ε is replaced by $F_\varepsilon F$, where F is any G-outer function satisfying $0 < \delta_1 \leqslant |F| \leqslant \delta_2 < \infty$. This somewhat mysterious outer function also appears in the description of the invariant subspaces of the Hardy space of a multiply connected domain [6, 7, 44, 78]. Although the F_ε's are not unique, we will show in Corollary 3.6.3 that the other parameters $\Theta_\mathcal{M}, \rho$, and E, are (essentially) unique.

1.3 Nearly invariant subspaces

The main tool in proving Theorem 1.2.3, which turns out to be interesting in its own right, is the concept of a 'nearly invariant' subspace first explored by Sarason [67] and Hitt [44]. Indeed, as we shall see in Corollary 6.1.3, if $\mathcal{M}$ is an invariant subspace of $H^2(G)$, then

$$\mathcal{N} := \left\{ f \circ \alpha^{-1} : f \in \mathcal{M} \right\},$$

where α is a particular conformal map from G onto $\widehat{\mathbb{C}} \setminus \gamma$ (see (2.3.8) below), is a nearly invariant subspace of $H^2(\widehat{\mathbb{C}} \setminus \gamma)$. Here $\widehat{\mathbb{C}} = \mathbb{C} \cup \{\infty\}$ is the Riemann sphere and

$$\gamma = \alpha(\partial G) = \left\{ e^{it} : -\frac{\pi}{2} \leqslant t \leqslant \pi \right\}.$$

We say a subspace $\mathcal{N} \subset H^2(\widehat{\mathbb{C}} \setminus \gamma)$, for which the origin is not a common zero for functions in $\mathcal{N}$, is *nearly invariant* if $f/z \in \mathcal{N}$ whenever $f \in \mathcal{N}$ and $f(0) = 0$. Nearly invariant subspaces have been, and continue to be, explored in various settings [5, 7, 39, 44, 54, 67]. In fact, very recently, nearly invariant subspaces have even appeared in mathematical physics [50].

If φ is the normalized reproducing kernel at the origin for $\mathcal{N}$, we will show in Corollary 3.3.1 that the operator

$$J : \mathcal{N} \to H^2(\mathbb{D}) \oplus L^2(\gamma, \omega), \quad Jf = \left(\frac{f_i}{\varphi_i}, f_e - f_i \frac{\varphi_e}{\varphi_i} \right)$$

is an isometry. Here ω is harmonic measure for $\widehat{\mathbb{C}} \setminus \gamma$ and, for $f \in H^2(\widehat{\mathbb{C}} \setminus \gamma)$,

$$f_i := f|\mathbb{D}, \quad f_e := f|\mathbb{D}_e.$$

In the expression,

$$f_e - f_i \frac{\varphi_e}{\varphi_i}$$

in the second component in the definition of Jf, we are using the appropriate almost everywhere defined boundary functions, i.e.,

$$f_e(\zeta) := \lim_{r \to 1^+} f(r\zeta), \quad f_i(\zeta) := \lim_{r \to 1^-} f(r\zeta), \quad \zeta \in \gamma.$$

If M_ζ denotes the operator on $H^2(\mathbb{D}) \oplus L^2(\gamma, \omega)$ defined by multiplication by the independent variable on each component function, then $(J\mathcal{N})^\perp$ becomes an invariant subspace for M_ζ and we use the Wold decomposition for $M_\zeta|(J\mathcal{N})^\perp$ to describe $(J\mathcal{N})^\perp$. We then use annihilators to describe $\mathcal{N}$ (Theorem 3.1.2). Our description of the nearly invariant subspaces of $H^2(\widehat{\mathbb{C}} \setminus \gamma)$ is the following.

Theorem 1.3.1. *Let $\mathcal{N}$ be a non-trivial nearly invariant subspace of $H^2(\widehat{\mathbb{C}} \setminus \gamma)$ with greatest common $\widehat{\mathbb{C}} \setminus \gamma$-inner divisor $\Theta_{\mathcal{N}}$. Then there exists a $\mathbb{D}$-outer function F, a measurable set $E \subset \gamma$, and a measurable function $\rho : \gamma \to \mathbb{C}$ such that*

$$\mathcal{N} = \Theta_{\mathcal{N}} \cdot \left\{ f \in H^2(\widehat{\mathbb{C}} \setminus \gamma) : \frac{f_i}{F} \in H^2(\mathbb{D}), f_i = \rho f_e \text{ a.e. on } E \right\}.$$

In Theorem 4.2.5 we describe the nearly invariant subspaces of $H^2(\widehat{\mathbb{C}} \setminus \gamma)$ in terms of the invariant subspaces of the backward shift operator

$$S^*f := \frac{f - f(0)}{z}$$

on $H^2(\mathbb{D})$. In Theorem 5.2.3 we describe the nearly invariant subspaces of $H^2(\widehat{\mathbb{C}} \setminus \gamma)$ in terms of de Branges-type spaces on $\widehat{\mathbb{C}} \setminus \mathbb{T}$.

1.4   Cyclic invariant subspaces

We also study the *cyclic* invariant subspaces of $H^2(G)$. By this we mean those invariant subspaces which take the form

$$[f] := \bigvee \{S^n f : n \in \mathbb{N}_0\}.$$

Here $\bigvee$ denotes the closed linear span in $H^2(G)$ and $\mathbb{N}_0 = \mathbb{N} \cup \{0\}$. Not every invariant subspace of $H^2(G)$ is cyclic. In fact, $H^2(G)$ is not a cyclic subspace [4, Cor. 3.3] (see Example 8.2.13 for other examples). We have the following result about cyclic invariant subspaces.

Theorem 1.4.1. *If $E \subset [0,1]$ is the measurable set corresponding to the (non-zero) invariant subspace $\mathcal{M}$ from Theorem 1.2.3 and $[0,1] \setminus E$ has positive measure then $\mathcal{M}$ is not cyclic.*

Though not every $\mathcal{M}$ is cyclic, we will show in Theorem 7.1.1 that every $\mathcal{M}$ is 2-cyclic.

Theorem 1.4.2. *For an invariant subspace $\mathcal{M}$ of $H^2(G)$, there are two functions $f, g \in \mathcal{M}$ so that*

$$\mathcal{M} = \bigvee \{z^m f, z^n g : m, n \in \mathbb{N}_0\}.$$

In fact, one can take f and g to be certain 'extremal' functions for $\mathcal{M}$. In Theorem 7.1.2 we will determine, for general $f, g \in H^2(G)$, when the invariant subspace generated by f and g is all of $H^2(G)$.

Theorem 1.4.3. *If $f, g \in H^2(G) \setminus \{0\}$, then*

$$\bigvee \{z^m f, z^n g : m, n \in \mathbb{N}_0\} = H^2(G)$$

if and only if f and g have no common G-inner factor and the set

$$\left\{ x \in [0,1) : \frac{f^+(x)}{f^-(x)} = \frac{g^+(x)}{g^-(x)} \right\}$$

has Lebesgue measure zero.

Though not every invariant subspace of $H^2(G)$ is cyclic, we can, under certain circumstances, compute the cyclic invariant subspace $[f]$ (see Theorem 7.2.2).

Theorem 1.4.4. *Suppose that both f and $1/f$ belong to $H^2(G)$ and $\rho := f^+/f^-$. Then $[f] = \mathcal{M}(\rho, [0,1])$.*

The equality $[f] = \mathcal{M}(\rho, [0,1])$ is not true in general (see Example 7.2.4).

In Corollary 6.1.3 we will show that every invariant subspace of $H^2(G)$ is nearly invariant. This implies, assuming $\mathcal{M} \neq \{0\}$, that for any $\lambda \in G$, $\dim(\mathcal{M} \ominus (z-\lambda)\mathcal{M}) = 1$. However, the analogue of Beurling's Theorem [1] does not hold in the slit disk, in the sense that $\mathcal{M} \ominus (z-\lambda)\mathcal{M}$ does not always generate $\mathcal{M}$. This is simply because not every invariant subspace of $H^2(G)$ ($H^2(G)$ in fact!) is cyclic (see also Example 8.2.13).

1.5 Essential spectrum

It is known that if $Sf = zf$ on $H^2(G)$, then $\sigma(S)$, the spectrum of S, is equal to $G^- = \mathbb{D}^-$ [20, Prop. 4.1] and that $\sigma_e(S)$, the essential spectrum of S, is equal to ∂G [20, Thm. 4.3]. In Theorem 8.2.5, we compute the essential spectrum of $S|\mathcal{M}$, where $\mathcal{M}$ is an invariant subspace of $H^2(G)$.

Theorem 1.5.1. *Let $\mathcal{M}$ be a non-zero invariant subspace of $H^2(G)$ and let $A(\mathcal{M})$ be the set of points $x \in [0,1)$ with the property that there exists an $f_x \in \mathcal{M}$ such that f/f_x extends to be analytic in a neighborhood of x whenever $f \in \mathcal{M}$. Then we have*

$$\sigma_e(S|\mathcal{M}) = \partial G \setminus A(\mathcal{M}).$$

Although every cyclic invariant subspace is contained in $\mathcal{M}(\rho, [0,1])$ for some ρ and Theorem 1.4.4 says that certain cyclic subspaces are of the form $\mathcal{M}(\rho, [0,1])$, Theorem 1.5.1 enables us to do the following (see Example 8.2.13).

Theorem 1.5.2. *There are measurable functions $\rho : [0,1] \to \mathbb{C}$ for which the invariant subspace $\mathcal{M}(\rho, [0,1])$ is not cyclic.*

As mentioned earlier, our results have analogs when the slit domain $G = \mathbb{D} \setminus [0,1]$ is replaced by a slit domain of the form in (1.1.1) (Theorem 10.1.2). Our results also have analogs when the Hardy space $H^2(G)$ is replaced by the Hardy-Smirnov space $E^2(G)$ (see (11.0.1)). Finally, we mention that in Theorem 7.3.2 we apply our main theorems to describe $P^2(\omega)$, where ω is harmonic measure on $G = \mathbb{D} \setminus [0,1)$ and $P^2(\omega)$ is the closure of the analytic polynomials in $L^2(\omega)$. Along the way, we compute the set of bounded point evaluations for $P^2(\omega)$.

---

[1] If one looks at the proof of Beurling's theorem from the Preamble, one can see, for a (non-zero) invariant subspace $\mathcal{M}$ of $H^2(\mathbb{D})$, that $\dim(\mathcal{M} \ominus z\mathcal{M}) = 1$ and that the invariant subspace generated by $\mathcal{M} \ominus z\mathcal{M}$ is $\mathcal{M}$.

Chapter 2

Preliminaries

2.1 Hardy space of a general domain

In this chapter, we set our notation and review some elementary facts about the Hardy spaces of general (simply connected) domains. Some good references for this material are [20, 22, 23, 31, 32]. For a simply connected domain $\Omega \subset \widehat{\mathbb{C}} := \mathbb{C} \cup \{\infty\}$, we say that an upper semicontinuous function $u : \Omega \to [-\infty, \infty)$ is *subharmonic* if it satisfies the *sub-mean value property*. That is to say, at each point $a \in \Omega$, there is an $r > 0$ so that

$$u(a) \leqslant \int_0^{2\pi} u(a + re^{i\theta}) \frac{d\theta}{2\pi}. \tag{2.1.1}$$

If f is analytic on Ω and $p > 0$, then $|f|^p$ is subharmonic. We say that a subharmonic function u has a *harmonic majorant* if there is a harmonic function U on Ω such that $u \leqslant U$ on Ω. By the Perron process for solving the classical Dirichlet problem [10, p. 200] [58, p. 118], one can show that if a subharmonic function $u \not\equiv -\infty$ has a harmonic majorant, then u has a *least harmonic majorant* U in that $u \leqslant U \leqslant V$ on Ω for all harmonic majorants V of u.

We say that an analytic function f on Ω belongs to the *Hardy space* $H^2(\Omega)$ if the subharmonic function $|f|^2$ has a harmonic majorant in Ω. If $z_0 \in \Omega$, we can norm $H^2(\Omega)$ by

$$\|f\|_{H^2(\Omega)} = \sqrt{U_f(z_0)}, \tag{2.1.2}$$

where U_f is the least harmonic majorant for $|f|^2$. By the mean value property for harmonic functions (i.e., equality in (2.1.1)[1]), notice that either $U_f > 0$ or $U_f \equiv 0$ on Ω. Thus $\|f\|_{H^2(\Omega)}$ actually defines a norm on $H^2(\Omega)$. Furthermore, we can use Harnack's inequality[2] to show that different norming points z_0 yield equivalent norms on $H^2(\Omega)$.

---

[1]In fact [18, p. 260], $u \in C(\Omega)$ is harmonic on Ω if and only if u satisfies the mean value property on Ω.

[2]Harnack's inequality: For fixed $z_1, z_2 \in \Omega$ there is a $C > 0$ so that $C^{-1}U(z_1) \leqslant U(z_2) \leqslant CU(z_1)$ for every positive harmonic function U on Ω [58, p. 14].

The following three simple facts will be used several times:

1. Suppose Ω_1, Ω_2 are two simply connected domains in $\widehat{\mathbb{C}}$ and ϕ is a conformal map from Ω_1 onto Ω_2. If $z_0 \in \Omega_1$ is the norming point for $H^2(\Omega_1)$ and $\phi(z_0)$ is the norming point for $H^2(\Omega_2)$, the operator

$$f \mapsto f \circ \phi \qquad (2.1.3)$$

is a unitary operator from $H^2(\Omega_2)$ onto $H^2(\Omega_1)$.

2. Let $\Omega_1 \subset \Omega_2$ and $z_0 \in \Omega_1$. Suppose z_0 is the norming point for both $H^2(\Omega_1)$ and $H^2(\Omega_2)$. If $f \in H^2(\Omega_2)$, then $f \in H^2(\Omega_1)$ and

$$\|f\|_{H^2(\Omega_1)} \leqslant \|f\|_{H^2(\Omega_2)}. \qquad (2.1.4)$$

3. Given a compact set $K \subset \Omega$, there is a positive constant C_K, depending only on K, such that

$$|f(z)| \leqslant C_K \|f\|_{H^2(\Omega)}, \quad z \in K, \quad f \in H^2(\Omega). \qquad (2.1.5)$$

In certain cases (see Proposition 2.4.13 below) one can estimate the constant C_K.

The inequality in (2.1.5) says that $H^2(\Omega)$, when endowed with the norm in (2.1.2), becomes a Hilbert space of analytic functions on Ω with inner product

$$\langle f, g \rangle = U_f(z_0)\overline{U_g(z_0)}.$$

By this we mean that $H^2(\Omega)$ is not only a Hilbert space comprised of analytic functions on Ω but the natural injection $i : H^2(\Omega) \to \mathrm{Hol}(\Omega)$ (the analytic functions on Ω endowed with the topology of uniform convergence on compact sets) is continuous. The inequality in (2.1.5) also says that for each fixed $z \in \Omega$, the linear functional $f \mapsto f(z)$ is continuous and so, by the Riesz representation theorem, there is a $k_z \in H^2(\Omega)$ such that

$$\langle f, k_z \rangle = f(z) \quad \forall f \in H^2(\Omega).$$

Thus $H^2(\Omega)$ is a *reproducing kernel Hilbert space*. We will see in a moment that the inner product $\langle f, g \rangle$ can be represented as an integral.

When Ω is the open unit disk $\mathbb{D}$, there is a more classical, but equivalent, definition of $H^2(\mathbb{D})$ in terms of integral means. Indeed, as mentioned in the Preamble at the beginning of this book, an analytic function f on $\mathbb{D}$ belongs to $H^2(\mathbb{D})$ if and only if

$$\sup_{0<r<1} \int_0^{2\pi} |f(re^{i\theta})|^2 \frac{d\theta}{2\pi} < \infty.$$

In fact, if we take the norming point in (2.1.2) for $H^2(\mathbb{D})$ to be $z_0 = 0$, we have

$$\|f\|_{H^2(\mathbb{D})}^2 = \sup_{0<r<1} \int_0^{2\pi} |f(re^{i\theta})|^2 \frac{d\theta}{2\pi}.$$

The classical theory of Hardy spaces [31] says that each $f \in H^2(\mathbb{D})$ has finite *radial limits*

$$f(e^{i\theta}) := \lim_{r \to 1^-} f(re^{i\theta}) \tag{2.1.6}$$

for almost every θ and

$$\|f\|^2_{H^2(\mathbb{D})} = \int_0^{2\pi} |f(e^{i\theta})|^2 \frac{d\theta}{2\pi}. \tag{2.1.7}$$

If

$$f(z) = \sum_{n=0}^{\infty} a_n z^n,$$

a simple computation will show that

$$\|f\|^2_{H^2(\mathbb{D})} = \sum_{n=0}^{\infty} |a_n|^2. \tag{2.1.8}$$

If the domain Ω is particularly nice, for example a Jordan domain with piecewise smooth boundary, the same sort of identity in (2.1.7) occurs except that arc length measure on the circle is replaced by harmonic measure ω_{z_0} (with norming point z_0) on $\partial\Omega$. See (2.2.7) below.

We will also discuss the *Hardy-Smirnov* class $E^2(\Omega)$ [31, 57, 72] of analytic functions f on Ω with the property that there exists a sequence $(\gamma_n)_{n \geqslant 1}$ of rectifiable Jordan curves which exhaust Ω (meaning that the curves eventually surround every compact subset of Ω) such that

$$\sup_n \int_{\gamma_n} |f|^2 ds < \infty.$$

In the above, ds is arc-length measure. A theorem of Keldysh and Lavrentiev ([31, p. 168] [46] [57, p. 146] [72]) says that $f \in E^2(\Omega)$ if and only if for any conformal map $\phi : \mathbb{D} \to \Omega$

$$\sup_{0 < r < 1} \int_{\phi(\{|z|=r\})} |f|^2 ds < \infty.$$

Since this last condition is independent of the choice of conformal map ϕ, we can set

$$\|f\|_{E^2(\Omega)} := \left(\sup_{0 < r < 1} \int_{\phi(\{|z|=r\})} |f|^2 ds \right)^{1/2}. \tag{2.1.9}$$

In other words,

$$f \in E^2(\Omega) \Leftrightarrow (f \circ \phi)(\phi')^{1/2} \in H^2(\mathbb{D}). \tag{2.1.10}$$

If $\psi := \phi^{-1}$, one can show that

$$H^2(\Omega) = (\psi')^{-1/2} E^2(\Omega).$$

In fact [31, p. 169] [75], $E^2(\Omega) = H^2(\Omega)$ if and only if $0 < m \leqslant |\psi'| \leqslant M < \infty$ on Ω. If Ω is bounded by an analytic curve, then, by basic properties of conformal maps, the above condition holds and so $H^2(\Omega) = E^2(\Omega)$. Roughly speaking, when $\partial\Omega$ has a 'corner' (as in the case of a slit domain), $H^2(\Omega) \neq E^2(\Omega)$. See below for a more detailed discussion of 'corners'. We direct the reader to [31, 47, 75] for a discussion of $E^2(\Omega)$ for multiply connected domains. These Hardy-Smirnov classes will appear both as a technical tool (see the proof of Theorem 7.2.2) as well as in E^2 versions of our main results for H^2 (see Chapter 11).

2.2   Harmonic measure

Some nice sources for this are [23, 35, 58]. We follow [58]. For a bounded Borel function $h : \partial\Omega \to \mathbb{R}$ let

$$(\mathcal{P}h)(z) := \sup_u u(z), \quad z \in \Omega,$$

where the supremum is taken over all subharmonic functions u on Ω such that

$$\limsup_{z \to \zeta} u(z) \leqslant h(\zeta) \quad \forall \zeta \in \partial\Omega.$$

The function $\mathcal{P}h$ is called the *Perron function* associated with h and is a bounded harmonic function on Ω. We are assuming that Ω is simply connected. If we also assume that $\widehat{\mathbb{C}} \setminus \Omega$ contains at least two points – which will definitely be the case here – then Ω is a *regular domain* in that for any $h \in C_{\mathbb{R}}(\partial\Omega)$, the continuous real-valued functions on $\partial\Omega$, we have

$$\lim_{z \to \zeta} (\mathcal{P}h)(z) = h(\zeta) \quad \forall \zeta \in \partial\Omega,$$

i.e., $\mathcal{P}h$ is the solution to the Dirichlet problem with boundary data h.

One can show, for a fixed point $z_0 \in \Omega$, that the map

$$h \mapsto (\mathcal{P}h)(z_0)$$

is a positive linear functional on $C_{\mathbb{R}}(\partial\Omega)$ of norm one and thus, by the Riesz representation theorem, there is a unique probability measure ω_{z_0} on $\partial\Omega$ such that

$$(\mathcal{P}h)(z_0) = \int_{\partial\Omega} h\, d\omega_{z_0}.$$

This unique measure ω_{z_0} is called the *harmonic measure* for Ω at z_0. When we need to emphasize the domain Ω, we will use the notation $\omega_{z_0,\Omega}$. When the point z_0 and the domain Ω are clear from context, we will drop these subscripts and use ω to denote $\omega_{z_0,\Omega}$.

From classical harmonic analysis, we know, for $h \in C_{\mathbb{R}}(\mathbb{T})$, that the function

$$z \mapsto \int_{\mathbb{T}} \frac{1 - |z|^2}{|\zeta - z|^2} h(\zeta) dm(\zeta),$$

where $dm = d\theta/2\pi$ is normalized Lebesgue measure on the unit circle $\mathbb{T} = \partial\mathbb{D}$, is harmonic on $\mathbb{D}$ and is the solution to the Dirichlet problem with boundary data h. Thus

$$d\omega_{z_0,\mathbb{D}} = \frac{1 - |z_0|^2}{|\zeta - z_0|^2} dm \qquad (2.2.1)$$

is the harmonic measure for $\mathbb{D}$ at $z_0 \in \mathbb{D}$. Observe that $d\omega_{0,\mathbb{D}} = dm$.

Here are a few basic facts about harmonic measure we will use several times:

1. For any z_0 and z_1 in Ω, the measures ω_{z_0} and ω_{z_1} are boundedly mutually absolutely continuous, i.e., there is a $C > 0$ such that

$$C\omega_{z_0}(E) \leqslant \omega_{z_1}(E) \leqslant \frac{1}{C}\omega_{z_0}(E)$$

for every Borel set $E \subset \partial\Omega$.

2. If Ω is a Jordan domain with piecewise analytic[3] boundary and $\phi : \mathbb{D} \to \Omega$ is an onto conformal map (with $\psi := \phi^{-1}$), Carathéodory's theorem [56, p. 18] says that ϕ extends to a homeomorphism from $\mathbb{D}^-$ onto Ω^-. Furthermore [56, p. 48, 52], ϕ' extends to be continuous and non-zero on $\mathbb{D}^-$ except for a finite number of points on $\mathbb{T}$, which turn out to be the inverse images under ϕ of the 'corners' of $\partial\Omega$. Changing variables and using the definition of harmonic measure shows that

$$\omega_{\phi(z_0),\Omega} = \omega_{z_0,\mathbb{D}} \circ \psi. \tag{2.2.2}$$

Now one can use (2.2.2) along with (2.2.1) and Jacobians to show that

$$d\omega_{\phi(0),\Omega} = |\psi'|\frac{ds}{2\pi}, \tag{2.2.3}$$

where ds is arc-length measure on $\partial\Omega$.

3. If $\phi(0)$ is replaced by some other point $z_0 \in \Omega$, we get

$$d\omega_{z_0,\Omega} = \frac{1}{2\pi}\frac{1 - |\psi(z_0)|^2}{|\psi(z) - \psi(z_0)|^2}|\psi'(z)|ds \asymp |\psi'|ds. \tag{2.2.4}$$

Let us make a few comments about estimates for the harmonic measure on our Jordan domain Ω with piecewise analytic boundary. We follow [4, Lemma 2.8]. If $\lambda \in \partial\Omega$ and $\partial\Omega$ is smooth near λ (i.e., λ is not one of the 'corners' of $\partial\Omega$), then ϕ not only extends to be continuous on $\mathbb{D}^-$ (via Carathéodory's theorem) but ϕ' extends to be continuous (and non-vanishing) up to $\mathbb{D} \cup J$, where J is some arc of the circle containing $\psi(\lambda)$ [56, p. 52]. So in this case,

$$d\omega_{z_0} \asymp ds$$

near λ. On the other hand, if $\lambda \in \partial\Omega$ is a corner with angle $\theta \in (0, \pi]$, then the map

$$z \mapsto (z - \lambda)^{\pi/\theta} \tag{2.2.5}$$

---

[3] Just to be clear, a curve Γ is *analytic* if it can be parameterized by a function $c(t)$ on $[a,b]$ such that c is an analytic function of t at every point of $[a,b]$ and $\dot{c}(t) \neq 0$ for all t. A curve Γ is *piecewise analytic* if it is the finite union of analytic curves. In all of what we say below, 'piecewise analytic' can be relaxed to 'piecewise Dini smooth'. See [56, p. 52] for a definition. In this monograph though, all of our boundary curves will be essentially piecewise analytic.

maps a region $R_\lambda \subset \Omega$ (see Figure 2.1) near λ to a domain with smooth boundary near the origin.

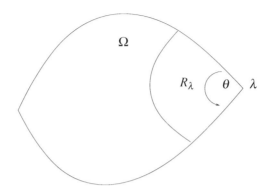

Figure 2.1: The piecewise analytic Jordan domain Ω with a corner at λ.

Hence the map

$$z \mapsto (\phi(z) - \lambda)^{\pi/\theta}$$

maps $\mathbb{D} \cap \psi(R_\lambda)$ to a domain with smooth boundary near the origin. By what was said above

$$\frac{1}{C} < \left| \frac{d}{dz} (\phi(z) - \lambda)^{\pi/\theta} \right| < C$$

near $\psi(\lambda)$. Thus we get

$$|\psi'(z)| \asymp |z - \lambda|^{\frac{\pi}{\theta} - 1}, \quad z \in R_\lambda, \tag{2.2.6}$$

and so using (2.2.4) we have

$$d\omega_{z_0} \asymp |z - \lambda|^{\frac{\pi}{\theta} - 1} ds$$

near the corner λ.

With Ω a Jordan domain with piecewise analytic boundary, each $f \in H^2(\Omega)$ has a finite non-tangential limit almost everywhere on $\partial\Omega$ and, analogous to (2.1.7) in the $H^2(\mathbb{D})$ case, we have

$$\|f\|^2_{H^2(\Omega)} = \int_{\partial\Omega} |f|^2 d\omega, \tag{2.2.7}$$

where $\omega = \omega_{z_0, \Omega}$ and z_0 is the norming point for $H^2(\Omega)$. If $f \in E^2(\Omega)$, then f also has a non-tangential limit almost everywhere on $\partial\Omega$ and

$$\|f\|^2_{E^2(\Omega)} = \int_{\partial\Omega} |f|^2 ds.$$

2.3 Slit domains

The domains considered in this monograph are not Jordan domains but slit domains and hence the formula in (2.2.7) needs to take into account which side of the slit one views the boundary function. There is a general theory about this involving the 'Martin boundary' [38]. However, since the slit domains we consider here are particularly simple, we can compute the norm directly in terms of the boundary function.

The first type of slit domain we consider is the slit disk

$$G = \mathbb{D} \setminus [0,1).$$

Using a sequence of standard conformal maps (see the Appendix), one can produce a conformal map

$$\phi_G : \mathbb{D} \to G \qquad (2.3.1)$$

such that ϕ_G maps the arc

$$J = \left\{ e^{i\theta} : 0 \leqslant \theta \leqslant \frac{\pi}{2} \right\}$$

to the 'top part' of the slit, while mapping

$$\bar{J} = \left\{ e^{-i\theta} : 0 \leqslant \theta \leqslant \frac{\pi}{2} \right\}$$

to the 'bottom part' of the slit (see Figure 2.2).

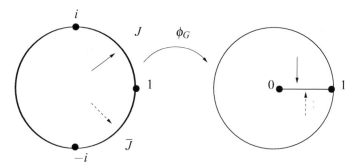

Figure 2.2: The conformal map $\phi_G : \mathbb{D} \to G$. Notice how the arc J gets mapped to the 'top' of the slit while the arc $\bar{J}$ gets mapped to the 'bottom' of the slit.

We are now ready to express the norm on $H^2(G)$ in terms of the boundary function. When $f \in H^2(G)$, recall from (2.1.3) that $f \circ \phi_G \in H^2(\mathbb{D})$ and so by conformal mapping and the fact that $f \circ \phi_G \in H^2(\mathbb{D})$ and thus has finite radial limits almost everywhere, we see that the limits

$$f^+(x) := \lim_{y \to 0^+} f(x+iy) \quad \text{and} \quad f^-(x) := \lim_{y \to 0^-} f(x+iy)$$

exist for almost every $x \in [0,1]$.

Moreover (see the Appendix), we have the identity

$$\phi_G(e^{i\theta}) = \overline{\phi_G(e^{-i\theta})}, \quad 0 \leqslant \theta \leqslant 2\pi. \tag{2.3.2}$$

From here we see that if $\psi_G = \phi_G^{-1}$, then

$$\left|(\psi_G')^+(x)\right| = \left|(\psi_G')^-(x)\right|, \quad 0 < x < 1. \tag{2.3.3}$$

Proposition 2.3.4. *For $f \in H^2(G)$,*

$$U_f(\phi_G(0)) = \|f\|^2_{H^2(G)} = \int_0^1 \left(|f^+|^2 + |f^-|^2\right) |\psi_G'| \frac{dx}{2\pi} + \int_0^{2\pi} |f|^2 |\psi_G'| \frac{d\theta}{2\pi},$$

where U_f is the least harmonic majorant of f and $|\psi_G'(x)|$ is the common value of $|(\psi_G')^+(x)|$ and $|(\psi_G')^-(x)|$ from (2.3.3).

Proof. For $f \in H^2(G)$,

$$\|f\|^2_{H^2(G)} = \|f \circ \phi_G\|^2_{H^2(\mathbb{D})}$$

$$= \int_J |f \circ \phi_G|^2 \frac{d\theta}{2\pi} + \int_{\tilde{J}} |f \circ \phi_G|^2 \frac{d\theta}{2\pi} + \int_{\pi/2}^{3\pi/2} |f \circ \phi_G|^2 \frac{d\theta}{2\pi}$$

$$= \int_0^1 |f^+|^2 |(\psi_G')^+| \frac{dx}{2\pi} + \int_0^1 |f^-|^2 |(\psi_G')^-| \frac{dx}{2\pi} + \int_0^{2\pi} |f|^2 |\psi_G'| \frac{d\theta}{2\pi}.$$

The result now follows from (2.3.3).   □

Since ϕ_G is not injective on parts of $\mathbb{T}$, ϕ_G no longer extends to be a homeomorphism between $\mathbb{D}^-$ and G^-. Thus the identity

$$\omega_{\phi_G(0),G} = \omega_{0,\mathbb{D}} \circ \psi$$

no longer makes sense and must be replaced by

$$\omega_{\phi_G(0),G} = \omega_{0,\mathbb{D}} \circ \psi^+ + \omega_{0,\mathbb{D}} \circ \psi^- + \omega_{0,\mathbb{D}} \circ \psi|\mathbb{T}.$$

By change of variables and using (2.3.3) we get

$$d\omega_{\phi_G(0),G} = \frac{1}{2\pi} |\psi'| dx + \frac{1}{2\pi} |\psi'| d\theta. \tag{2.3.5}$$

An analysis of $|\psi_G'|$ (see (2.3.16) below) will show that

$$d\omega_{\phi_G(0),G} \asymp |\xi|^{-1/2} |\xi - 1| ds. \tag{2.3.6}$$

Another type of slit domain we will consider in Chapter 3 is

$$\widehat{\mathbb{C}} \setminus \gamma, \quad \gamma := \left\{ e^{it} : -\frac{\pi}{2} \leqslant t \leqslant \pi \right\}. \tag{2.3.7}$$

Consider the conformal map

$$\alpha(z) := \frac{(1+z)^2(1-z)^{-2} - i}{(1+z)^2(1-z)^{-2} + i}. \tag{2.3.8}$$

A routine exercise with composition of several standard conformal maps (see the Appendix) shows that α maps G onto $\widehat{\mathbb{C}} \setminus \gamma$ and $\alpha(i(1 - \sqrt{2})) = \infty$. Also notice that

$$\alpha(\mathbb{T}) = \gamma := \{e^{it} : 0 \leqslant t \leqslant \pi\},$$

$$\alpha([0,1]) = \gamma' := \left\{ e^{it} : \frac{3\pi}{2} \leqslant t \leqslant 2\pi \right\},$$

$$\alpha([-1,0]) = \gamma''' := \left\{ e^{it} : \pi \leqslant t \leqslant \frac{3\pi}{2} \right\}$$

(see Figure 2.3). Furthermore, if $\mathbb{D}_+ := \mathbb{D} \cap \{\Im z > 0\}$ and $\mathbb{D}_- := \mathbb{D} \cap \{\Im z < 0\}$, then

$$\alpha(\mathbb{D}_-) = \mathbb{D}_e := \widehat{\mathbb{C}} \setminus \mathbb{D}^-, \quad \alpha(\mathbb{D}_+) = \mathbb{D}$$

(see Figure 2.4).

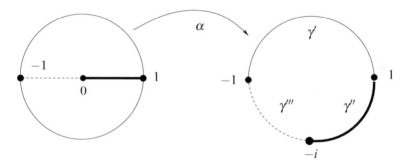

Figure 2.3: The conformal map $\alpha : G \to \widehat{\mathbb{C}} \setminus \gamma$. Note that $\gamma = \alpha(\mathbb{T})$, $\gamma'' = \alpha([0,1])$, and $\gamma''' = \alpha([-1,0])$.

Finally, we have the identity

$$\alpha(e^{i\theta}) = \alpha(e^{-i\theta}), \quad 0 \leqslant \theta \leqslant 2\pi. \tag{2.3.9}$$

For $f \in H^2(\widehat{\mathbb{C}} \setminus \gamma)$, define

$$f_i(z) := f(z), \quad z \in \mathbb{D} \quad \text{and} \quad f_e(z) := f(z), \quad z \in \mathbb{D}_e. \tag{2.3.10}$$

If $f \in H^2(\widehat{\mathbb{C}} \setminus \gamma)$, use the conformal map

$$\phi_\gamma := \alpha \circ \phi_G : \mathbb{D} \to \widehat{\mathbb{C}} \setminus \gamma \tag{2.3.11}$$

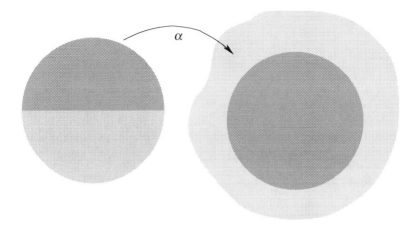

Figure 2.4: The conformal map $\alpha : G \to \widehat{\mathbb{C}} \setminus \gamma$. Note that $\alpha(\mathbb{D}_+) = \mathbb{D}$ and $\alpha(\mathbb{D}_-) = \mathbb{D}_e$.

and the fact that $f \circ \phi_\gamma$ belongs to $H^2(\mathbb{D})$ and thus has radial limits almost everywhere to see that the limits

$$f_e(e^{i\theta}) := \lim_{r \to 1^+} f(re^{i\theta}) \quad \text{and} \quad f_i(e^{i\theta}) := \lim_{r \to 1^-} f(re^{i\theta})$$

exist almost everywhere. Use the proof of Proposition 2.3.4, along with the identity

$$\phi_\gamma(e^{-i\theta}) = \phi_\gamma(e^{i\theta}), \quad 0 \leqslant \theta \leqslant 2\pi,$$

(which comes from (2.3.2) and (2.3.9)), and the above facts about the conformal maps ϕ_G and α, to prove the following.

Proposition 2.3.12. *For* $f \in H^2(\widehat{\mathbb{C}} \setminus \gamma)$,

$$U_f(\phi_\gamma(0)) = \|f\|^2_{H^2(\widehat{\mathbb{C}} \setminus \gamma)} = \int_{-\pi/2}^{\pi} \left(|f_i|^2 + |f_e|^2 \right) |\psi_\gamma'| \frac{d\theta}{2\pi},$$

where $\psi_\gamma := \phi_\gamma^{-1}$.

By (2.3.16) one can show that

$$d\omega_{\phi_\gamma(0),\widehat{\mathbb{C}} \setminus \gamma} = |\psi_\gamma'| \frac{d\theta}{2\pi} \asymp |\xi + 1|^{-1/2} |\xi + i|^{-1/2} ds. \qquad (2.3.13)$$

The next types of domains we consider in Chapter 5 are a certain subclass of domains of the form

$$\Omega := \mathbb{D} \setminus \Gamma,$$

where Γ is a simple analytic curve which meets $\mathbb{T}$ at a positive angle (see Figure 2.5). Let μ be the endpoint of Γ in $\mathbb{D}$ and λ the endpoint of Γ in $\mathbb{T}$ and θ be the *acute* angle between Γ and $\mathbb{T}$ at λ.

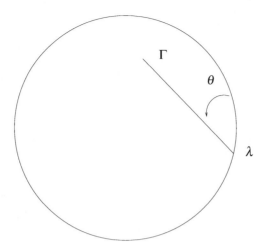

Figure 2.5: The slit domain $\Omega = \mathbb{D} \setminus \Gamma$.

An analysis similar to the one used to derive the estimate in (2.2.6), but with the domains R_λ replaced separately by domains R_λ^+ (near the top of the slit) and R_λ^- (near the bottom of the slit) and away from the other endpoint μ (see Figure 2.6), will yield the estimate

$$|\psi'(z)| \asymp \begin{cases} |z - \lambda|^{\frac{\pi}{\theta} - 1}, & z \in R_\lambda^+; \\ |z - \lambda|^{\frac{\pi}{\pi - \theta} - 1}, & z \in R_\lambda^-. \end{cases} \tag{2.3.14}$$

where $\phi : \mathbb{D} \to \Omega$ and $\psi := \phi^{-1}$.

To get an estimate of $|\psi'|$ near μ (the endpoint of Γ in $\mathbb{D}$) we replace the map in (2.2.5) with the map $z \mapsto (z - \mu)^{1/2}$ and do the same analysis as before to get

$$|\psi'| \asymp |z - \mu|^{-1/2} \tag{2.3.15}$$

when z is near μ. Combining (2.3.14) and (2.3.15) we get

$$|(\psi')^+(\xi)| \asymp |\xi - \mu|^{-1/2}|\xi - \lambda|^{\frac{\pi}{\theta} - 1}; \tag{2.3.16}$$
$$|(\psi')^-(\xi)| \asymp |\xi - \mu|^{-1/2}|\xi - \lambda|^{\frac{\pi}{\pi - \theta} - 1}.$$

The identity

$$\omega_{\phi(0),\Omega} = \omega_{0,\mathbb{D}} \circ \psi^+ + \omega_{0,\mathbb{D}} \circ \psi^- + \omega_{0,\mathbb{D}} \circ \psi|\mathbb{T}$$

yields

$$d\omega_{\phi(0),\Omega} = |(\psi')^+|\frac{ds}{2\pi} + |(\psi')^-|\frac{ds}{2\pi} + |\psi'|\mathbb{T}\frac{d\theta}{2\pi}$$

from which follows an estimate of harmonic measure near the corners of $\partial\Omega$. If the angle θ is $\pi/2$ then $|(\psi')^+(\xi)| \asymp |(\psi')^-(\xi)|$. When $\theta \neq \pi/2$, $|(\psi')^+(\xi)|$ is considerably different than $|(\psi')^-(\xi)|$.

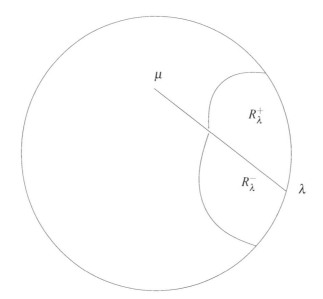

Figure 2.6: The regions R_λ^+ and R_λ^-.

Following the proof of Proposition 2.3.4 we get that

$$U_f(\phi(0)) = \|f\|_{H^2(\Omega)}^2 = \int_\Gamma |f^+|^2 |(\psi')^+| \frac{ds}{2\pi} + \int_\Gamma |f^-|^2 |(\psi')^-| \frac{ds}{2\pi} + \int_\mathbb{T} |f|^2 |\psi'| \frac{d\theta}{2\pi} \tag{2.3.17}$$

and remark that if $\theta \neq \pi/2$ the first two integrals can not be combined and so the identity in (2.2.7) is no longer valid but must be interpreted as in the identity in (2.3.17).

2.4   More about the Hardy space

If $\Omega \subset \widehat{\mathbb{C}}$ is either a piecewise analytic Jordan domain or one of the slit domains ($\mathbb{D} \setminus \Gamma$ or $\widehat{\mathbb{C}} \setminus \gamma$) considered previously, let ϕ be a conformal map from $\mathbb{D}$ onto Ω and $\psi := \phi^{-1}$. An analytic function f on Ω is *Ω-inner* if $f \circ \phi$ is inner in the classical sense, i.e., $|f \circ \phi(\zeta)| = 1$ for almost every $\zeta \in \mathbb{T}$, and *Ω-outer* if $f \circ \phi$ is outer in the classical sense, i.e.,

$$\log|f \circ \phi(0)| = \int_\mathbb{T} \log|f \circ \phi(\zeta)| dm(\zeta).$$

It is well known [31, p. 24] that every $f \in H^2(\mathbb{D})$ can be factored, uniquely up to a unimodular constant, as $f = IF$, where I is $\mathbb{D}$-inner and F is $\mathbb{D}$-outer. By conformal mapping, we see the following.

Proposition 2.4.1. *Every $f \in H^2(\Omega)$ can be factored uniquely, up to a unimodular constant, as $f = IF$, where I is Ω-inner and F is Ω-outer.*

For $H^2(\Omega)$ or $E^p(\Omega)$ of a multiply connected domain, factorization is more involved [47, 75].

If $f \in H^2(\mathbb{D}) \setminus \{0\}$, a classical theorem of Riesz [31, p. 17] says that

$$\log|f| \in L^1(\mathbb{T}, m). \tag{2.4.2}$$

By a change of variables, there is an analog of this for $H^2(\Omega)$: if $f \in H^2(\Omega) \setminus \{0\}$, then

$$\log|f| \in L^1(\partial\Omega, \omega^*). \tag{2.4.3}$$

This notation requires some explanation. When Ω is a piecewise analytic Jordan curve, (2.4.3) means

$$\int_{\partial\Omega} |\log|f|| d\omega < \infty.$$

When $\Omega = \mathbb{D} \setminus \Gamma$, (2.4.3) means

$$\int_{\Gamma} |\log|f^+|||(\psi')^+| \frac{ds}{2\pi} + \int_{\Gamma} |\log|f^-|||(\psi')^-| \frac{ds}{2\pi} + \int_{\mathbb{T}} |\log|f|||\psi'| \frac{d\theta}{2\pi}$$

is finite. When $\Omega = \widehat{\mathbb{C}} \setminus \gamma$, (2.4.3) means

$$\int_{\gamma} (|\log|f_i|| + |\log|f_e||)|\psi_\gamma'| \frac{d\theta}{2\pi}$$

is finite.

Define the *Nevanlinna class*

$$N(\Omega) := \left\{ \frac{f}{g} : f, g \in H^\infty(\Omega), g \not\equiv 0 \right\}$$

and the *Smirnov class*

$$N^+(\Omega) := \left\{ \frac{f}{g} : f, g \in H^\infty(\Omega), g \text{ is } \Omega\text{-outer} \right\}. \tag{2.4.4}$$

Using Beurling's theorem and conformal mapping (see Proposition 6.1.2 below) it follows that $f \in H^\infty(\Omega)$ is Ω-outer if and only if

$$\text{clos}_{H^2(\Omega)}\{hf : h \in H^\infty(\Omega)\} = H^2(\Omega).$$

From here we get the following (see the proof of Lemma 3.4.12).

Proposition 2.4.5. *Suppose Ω_1, Ω_2 are simply connected regions and $\Omega_1 \subset \Omega_2$. If $f \in H^\infty(\Omega_2)$ is Ω_2-outer, then $f|\Omega_1$ is Ω_1-outer and consequently we have the containment $N^+(\Omega_2)|\Omega_1 \subset N^+(\Omega_1)$.*

We will make many uses of this next theorem due to Polubarinova and Kochina [57, p. 80] (see also [31, p. 28]). For $f \in N^+(\mathbb{D})$ let $f|\mathbb{T}$ denote the non-tangential boundary function of f – which exists almost everywhere.

Theorem 2.4.6. *Let $0 < p < \infty$ and $f \in N^+(\mathbb{D})$. If $f|\mathbb{T} \in L^p(\mathbb{T}, m)$, then $f \in H^p(\mathbb{D})$.*

For $0 < p < \infty$, define the Hardy class

$$H^p(\Omega) := \{ f \in \text{Hol}(\Omega) : |f|^p \text{ has a harmonic majorant on } \Omega \} \qquad (2.4.7)$$

and the Hardy-Smirnov class,

$$E^p(\Omega) := \left\{ f \in \text{Hol}(\Omega) : (f \circ \phi)(\phi')^{1/p} \in H^p(\mathbb{D}) \right\} \qquad (2.4.8)$$

$$= \left\{ f \in \text{Hol}(\Omega) : \sup_{0 < r < 1} \int_{\phi(\{|z|=r\})} |f(z)|^p \, ds < \infty \right\},$$

where ϕ is a conformal map from $\mathbb{D}$ onto Ω. As mentioned earlier, we have

$$H^p(\Omega) = (\psi')^{-1/p} E^p(\Omega) \qquad (2.4.9)$$

and

$$H^p(\Omega) = E^p(\Omega) \Leftrightarrow 0 < c \leqslant |\psi'| \leqslant C < \infty \text{ on } \Omega.$$

Functions in $H^p(\Omega)$ and $E^p(\Omega)$ have finite non-tangential limits almost everywhere with respect to arc-length measure and the boundary function belongs to $L^p(\partial\Omega, \omega^*)$ – when the function belongs to $H^p(\Omega)$ - or $L^p(\partial\Omega, ds)$ – when the function belongs to $E^p(\Omega)$. As before (see (2.4.3)), when Ω is a piecewise analytic Jordan domain, the condition $f|\partial\Omega \in L^p(\Omega, \omega^*)$ means $f|\partial\Omega \in L^p(\partial\Omega, \omega)$. When $\Omega = \mathbb{D} \setminus \Gamma$, we mean

$$\int_\Gamma |f^+|^p |(\psi')^+| \frac{ds}{2\pi} + \int_\Gamma |f^-|^p |(\psi')^-| \frac{ds}{2\pi} + \int_\mathbb{T} |f|^p |\psi'| \frac{d\theta}{2\pi} < \infty.$$

When $\Omega = \widehat{\mathbb{C}} \setminus \gamma$ we mean

$$\int_\gamma (|f_i|^p + |f_e|^p) |\psi_\gamma'| \frac{d\theta}{2\pi} < \infty.$$

By a change of variables, one can prove the following generalization of Theorem 2.4.6.

Proposition 2.4.10. *Suppose $0 < p < \infty$ and $f \in N^+(\Omega)$.*

1. If $f|\partial\Omega \in L^p(\partial\Omega, \omega^*)$, then $f \in H^p(\Omega)$.

2. If $f|\partial\Omega \in L^p(\partial\Omega, ds)$,[4] then $f \in E^p(\Omega)$.

For $f \in H^1(\mathbb{D})$, there is the following version of the Cauchy integral formula [31, p. 41]:

$$f(z) = \frac{1}{2\pi i} \oint_\mathbb{T} \frac{f(\zeta)}{\zeta - z} d\zeta, \quad z \in \mathbb{D}. \qquad (2.4.11)$$

There is also a generalization of this formula for $E^1(\Omega)$ [31, p. 170].

---

[4]When Ω is a slit domain, we need to take into account f^+ and f^- (or f_i and f_e) in the above integral.

Proposition 2.4.12. *For $f \in E^1(\Omega)$,*

$$f(z) = \frac{1}{2\pi i} \oint_{\partial\Omega} \frac{f(\xi)}{\xi - z} d\xi, \quad z \in \Omega.$$

When Ω is a slit domain, we need to take into account f^+ and f^- (or f_i and f_e) in the above integral. We can now use the identity $H^p(\Omega) = (\psi')^{-1/p} E^p(\Omega)$ along with the Cauchy integral formula in Proposition 2.4.12 and a trick in [31, p. 36] involving inner-outer factorizations to establish the following pointwise estimates.

Proposition 2.4.13. *If $f \in E^p(\Omega)$, then*

$$|f(\lambda)| \leqslant C \frac{\|f\|_{E^p(\Omega)}}{(dist(\lambda, \partial\Omega))^{1/p}}, \quad \lambda \in \Omega.$$

If $f \in H^p(\Omega)$, then

$$|f(\lambda)| \leqslant C \frac{\|f\|_{H^p(\Omega)}}{|\psi'(\lambda)|^{1/p} (dist(\lambda, \partial\Omega))^{1/p}}, \quad \lambda \in \Omega.$$

The above proposition implies that functions in the unit ball of $H^p(\Omega)$ or $E^p(\Omega)$ form a normal family on Ω.

Chapter 3

Nearly invariant subspaces

3.1 Statement of the main result

For a general domain $\Omega \subset \widehat{\mathbb{C}}$, a subspace[1] $\mathcal{N} \subset H^2(\Omega)$ (not necessarily invariant) is said to be *nearly invariant* if there is some $\lambda_0 \in \Omega$ such that whenever $f \in \mathcal{N}$ and $f(\lambda_0) = 0$, then

$$\frac{f}{z - \lambda_0} \in \mathcal{N}.$$

The following useful proposition, found in [7, Prop. 5.1], says that we need not single out a particular λ_0.

Proposition 3.1.1. *If $\mathcal{N}$ is a subspace of $H^2(\Omega)$, then the following are equivalent.*

1. *$\mathcal{N}$ is nearly invariant.*

2. *If $Z(\mathcal{N})$ is the set of common zeros[2] of functions in $\mathcal{N}$, then*

$$\frac{f}{z - \lambda} \in \mathcal{N}$$

   whenever $f \in \mathcal{N}$ and $\lambda \in \Omega \setminus Z(\mathcal{N})$ with $f(\lambda) = 0$.

3. *For any $f, g \in \mathcal{N}$,*

$$\frac{f - \dfrac{f}{g}(\lambda)g}{z - \lambda} \in \mathcal{N}$$

   whenever $\lambda \in \Omega \setminus Z(\mathcal{N})$ and $g(\lambda) \neq 0$.

Nearly invariant subspaces of $H^2(\mathbb{D})$ were first explored by Sarason and Hitt [44, 67]. They proved the following theorem: Let $\mathcal{N}$ be a nearly invariant subspace of $H^2(\mathbb{D})$

---

[1] Recall from the introduction that a subspace will be a *closed* linear manifold.
[2] i.e., $Z(\mathcal{N}) := \bigcap_{f \in \mathcal{N}} f^{-1}(\{0\})$.

and let $f_{\mathcal{N}}$ be the unique function of unit norm in $\mathcal{N}$ which maximizes $\Re f(0)$. Then there is a unique $\mathbb{D}$-inner function Θ such that

$$\mathcal{N} = f_{\mathcal{N}} \cdot \left(H^2(\mathbb{D}) \ominus \Theta H^2(\mathbb{D}) \right).$$

More recently, nearly invariant subspaces have appeared in the study of invariant subspaces of $H^p(\Omega)$ for regions Ω with 'holes' [7, 44], surjective Toeplitz operators on $H^2(\mathbb{D})$ [36], inverse spectral theory [54], derivation invariant subspaces of C^∞ [5], and mathematical physics [50].

Our main theorem of this chapter characterizes the nearly invariant subspaces of $H^2(\widehat{\mathbb{C}} \setminus \gamma)$. Recall from (2.3.7) that

$$\gamma := \left\{ e^{it} : -\frac{\pi}{2} \leqslant t \leqslant \pi \right\}.$$

Theorem 3.1.2. *Let $\mathcal{N}$ be a non-trivial nearly invariant subspace of $H^2(\widehat{\mathbb{C}} \setminus \gamma)$ with greatest common $\widehat{\mathbb{C}} \setminus \gamma$-inner divisor $\Theta_{\mathcal{N}}$. Then there exists a $\mathbb{D}$-outer function F, a measurable set $E \subset \gamma$, and a measurable function $\rho : \gamma \to \mathbb{C}$ such that*

$$\mathcal{N} = \Theta_{\mathcal{N}} \cdot \left\{ f \in H^2(\widehat{\mathbb{C}} \setminus \gamma) : \frac{f_i}{F} \in H^2(\mathbb{D}), f_i = \rho f_e \text{ a.e. on } E \right\}. \qquad (3.1.3)$$

Remark 3.1.4.   1. The functions f_i and f_e were defined in (2.3.10).

2. Theorem 3.1.2 is still valid when the arc $\gamma = \{ e^{it} : -\pi/2 \leqslant t \leqslant \pi \}$ is replaced by any proper sub-arc of $\mathbb{T}$.

3. It is easy to check that when the linear manifold defined by the right-hand side of (3.1.3) is closed, it is a nearly invariant subspace. We also remark, as discussed following the statement of Theorem 1.2.3, that Theorem 3.1.2 does not say that for any $\mathbb{D}$-outer function F, the linear manifold defined in (3.1.3) is closed. Rather, it says that for a given nearly invariant subspace $\mathcal{N}$, there is a $\mathbb{D}$-outer function F for which the linear manifold in (3.1.3) is closed, i.e., a subspace.

The rest of this rather long chapter will be devoted to the proof of Theorem 3.1.2. In Chapter 6, we will use this theorem, together with a conformal mapping trick, to characterize the invariant subspaces of the slit domain $G = \mathbb{D} \setminus [0,1)$ (see Corollary 6.1.6 and Theorem 6.2.1).

3.2   Normalized reproducing kernels

In what follows, $\mathcal{N}$ will be a non-trivial nearly invariant subspace of $H^2(\widehat{\mathbb{C}} \setminus \gamma)$ such that the points $z = 0$ and $z = \infty$ are not in the common zero set of $\mathcal{N}$ [3] and φ will be the

---

[3]If this is not the case, one can always consider the nearly invariant subspace $\mathcal{N} := \mathcal{N}/B \circ \phi_\gamma^{-1}$ where ϕ_γ is a conformal map from $\mathbb{D}$ onto $\widehat{\mathbb{C}} \setminus \gamma$ and B is the $\mathbb{D}$-Blaschke product with zeros at $\phi_\gamma^{-1}(0)$ and $\phi_\gamma^{-1}(\infty)$ of appropriate orders.

normalized reproducing kernel in $\mathcal{N}$ at the origin. More precisely, the linear functional on $\mathcal{N}$ defined by $f \mapsto f(0)$ is continuous (and non-zero) and so, by the Riesz representation theorem, there is a $k_0^{\mathcal{N}} \in \mathcal{N}$ such that

$$f(0) = \langle f, k_0^{\mathcal{N}} \rangle \quad \forall f \in \mathcal{N}.$$

If we define

$$\varphi := \frac{k_0^{\mathcal{N}}}{\|k_0^{\mathcal{N}}\|},$$

then $\varphi \in \mathcal{N}$ and satisfies the two identities

$$\frac{f}{\varphi}(0) = \langle f, \varphi \rangle \quad \forall f \in \mathcal{N}, \tag{3.2.1}$$

$$\langle \varphi, \varphi \rangle = 1. \tag{3.2.2}$$

This normalized reproducing kernel φ is often called the 'extremal function' for $\mathcal{N}$ since it is the unique solution to the extremal problem

$$\sup \left\{ \Re g(0) : g \in \text{ball}(\mathcal{N}) \right\}.$$

One proves this by a variational argument [41]. Extremal functions have played an important role in studying invariant subspaces in various settings [7, 8, 27, 28, 29, 41, 48, 60, 61, 62, 73] and they will play an important role here.

We will also need the normalized reproducing kernel function for $\mathcal{N}$ at ∞,

$$\psi := \frac{k_\infty^{\mathcal{N}}}{\|k_\infty^{\mathcal{N}}\|}.$$

Here we have

$$\frac{f}{\psi}(\infty) = \langle f, \psi \rangle \quad \forall f \in \mathcal{N}, \tag{3.2.3}$$

$$\langle \psi, \psi \rangle = 1.$$

By the definition of nearly invariant, (2.1.5), and the closed graph theorem, one can show that the operator $L : \mathcal{N} \to \mathcal{N}$

$$Lf := \frac{f - \frac{f}{\varphi}(0)\varphi}{z} = \frac{f - \langle f, \varphi \rangle \varphi}{z} \tag{3.2.4}$$

is bounded. There is also the companion operator $R : \mathcal{N} \to \mathcal{N}$

$$Rf := z \left(f - \frac{f}{\psi}(\infty)\psi \right) = z(f - \langle f, \psi \rangle \psi). \tag{3.2.5}$$

Before discussing some specific properties of the operator L, there is the following nice relationship between R and L.

Theorem 3.2.6. *Let $\mathcal{N}$ be a non-trivial nearly invariant subspace of $H^2(\widehat{\mathbb{C}} \setminus \gamma)$ and let φ and ψ be the extremal functions for $\mathcal{N}$ considered above. Then we have the following.*

1. $L^* = R$.

2. *If $\lambda \in \widehat{\mathbb{C}} \setminus \gamma$ and $k_\lambda^{\mathcal{N}}$ is the reproducing kernel function for $\mathcal{N}$ at λ, then*

$$k_\lambda^{\mathcal{N}}(z) = \frac{\overline{\varphi(\lambda)}\varphi(z) - \overline{\lambda} z \overline{\psi(\lambda)}\psi(z)}{1 - \overline{\lambda} z}, \quad z \neq 1/\overline{\lambda}. \tag{3.2.7}$$

Proof. Since $\langle Lf, \psi \rangle = (Lf)(\infty) = 0$ for all $f \in \mathcal{N}$, we can say, for any $f, g \in \mathcal{N}$, that

$$\langle Lf, g \rangle = \left\langle Lf, g - \frac{g}{\psi}(\infty)\psi \right\rangle. \tag{3.2.8}$$

Since the inner product on $H^2(\widehat{\mathbb{C}} \setminus \gamma)$ is given by

$$\langle f, g \rangle = \int_{-\pi/2}^{\pi} (f_i \overline{g_i} + f_e \overline{g_e}) |\psi_\gamma'| \frac{d\theta}{2\pi}$$

(Proposition 2.3.12), we can use the fact that $\overline{z} = 1/z$ for $z \in \mathbb{T}$ to see that the quantity on the right-hand side of (3.2.8) is equal to

$$\left\langle f - \frac{f}{\varphi}(0)\varphi, z\left(g - \frac{g}{\psi}(\infty)\psi\right) \right\rangle = \left\langle f - \frac{f}{\varphi}(0)\varphi, Rg \right\rangle = \langle f, Rg \rangle.$$

The last equality follows since

$$\langle Rg, \varphi \rangle = \frac{Rg}{\varphi}(0) = 0.$$

This shows that $L^* = R$ which proves statement (1).

To prove statement (2), we compute $L^* k_\lambda^{\mathcal{N}}$ in two different ways. From the definition of L and the reproducing property of $k_\lambda^{\mathcal{N}}$ we have

$$\langle Lf, k_\lambda^{\mathcal{N}} \rangle = \frac{f(\lambda) - \langle f, \varphi \rangle \varphi(\lambda)}{\lambda} \quad \forall f \in \mathcal{N},$$

and so, replacing f with a reproducing kernel $k_z^{\mathcal{N}} \in \mathcal{N}$ and using the trivial identity $\overline{k_z^{\mathcal{N}}(\lambda)} = k_\lambda^{\mathcal{N}}(z)$, we have

$$(L^* k_\lambda^{\mathcal{N}})(z) = \overline{\langle k_z^{\mathcal{N}}, L^* k_\lambda^{\mathcal{N}} \rangle} = \overline{\langle Lk_z^{\mathcal{N}}, k_\lambda^{\mathcal{N}} \rangle} = \frac{k_\lambda^{\mathcal{N}}(z) - \overline{\varphi(\lambda)}\varphi(z)}{\overline{\lambda}}.$$

On the other hand, from the identity $L^* = R$, just shown above, we have

$$L^* k_\lambda^{\mathcal{N}} = z\left(k_\lambda^{\mathcal{N}} - \overline{\psi(\lambda)}\psi\right).$$

The formula in (3.2.7) now follows by equating the above two formulas for $L^* k_\lambda^{\mathcal{N}}$ and solving for $k_\lambda^{\mathcal{N}}$. $\qquad\square$

Corollary 3.2.9. *For each* $\lambda \in \widehat{\mathbb{C}} \setminus \gamma$ *we have the following.*

1.

$$\overline{\varphi(\lambda)}\varphi(1/\overline{\lambda}) = \overline{\psi(\lambda)}\psi(1/\overline{\lambda}).$$

2. *If* $|\lambda| < 1$, *then*

$$|\lambda||\psi(\lambda)| \leqslant |\varphi(\lambda)|$$

while if $|\lambda| > 1$, *then*

$$|\lambda||\psi(\lambda)| \geqslant |\varphi(\lambda)|.$$

3.

$$(k_\lambda^{\mathcal{N}}/\varphi)_i \in H^\infty(\mathbb{D}) \quad while \quad (k_\lambda^{\mathcal{N}}/\psi)_e \in H^\infty(\mathbb{D}_e).$$

4. $\mathcal{N}$ *is the smallest nearly invariant subspace of* $H^2(\widehat{\mathbb{C}} \setminus \gamma)$ *containing* φ *and* ψ.

Proof. The proof of statement (1) follows from the fact that since $k_\lambda^{\mathcal{N}}$ is analytic at $1/\overline{\lambda}$, the numerator in the formula for $k_\lambda^{\mathcal{N}}$ – see (3.2.7) – must vanish when $z = 1/\overline{\lambda}$. Statement (2) follows from (3.2.7) and the fact that $k_\lambda^{\mathcal{N}}(\lambda) = \|k_\lambda^{\mathcal{N}}\|^2 \geqslant 0$. Statement (3) follows from statement (2) and (3.2.7). Finally, to prove statement (4), use (3.2.7) to prove the identity

$$k_\lambda^{\mathcal{N}} = \frac{\overline{\varphi(\lambda)}\varphi - \overline{\psi(\lambda)}\psi}{1 - \overline{\lambda}z} + \overline{\psi(\lambda)}\psi.$$

Certainly the second summand above belongs to the nearly invariant subspace containing φ and ψ. As long as $1/\overline{\lambda}$ is not a common zero for $\mathcal{N}$, the first summand belongs to the nearly invariant subspace containing φ and ψ since the numerator belongs to this space and, via statement (1), the numerator vanishes at $z = 1/\overline{\lambda}$. The result now follows since finite linear combinations of the reproducing kernels $\{k_\lambda^{\mathcal{N}} : 1/\overline{\lambda} \notin Z(\mathcal{N})\}$ are dense in $\mathcal{N}$. □

We now focus on properties of the operator L defined in (3.2.4).

Proposition 3.2.10. *The operator* L *in* (3.2.4) *is a contraction on* $\mathcal{N}$.

Proof. For $f \in \mathcal{N}$ we can use (3.2.1) to get

$$f = \frac{f}{\varphi}(0)\varphi + zLf = \langle f, \varphi \rangle \varphi + zLf. \qquad (3.2.11)$$

This shows, by rearranging the terms in the above equation, that $zLf \in \mathcal{N}$ and moreover, by (3.2.1) and (3.2.2),

$$\langle zLf, \varphi \rangle = \langle f - \langle f, \varphi \rangle \varphi, \varphi \rangle = 0$$

and so $\varphi \perp zLf$. This yields

$$\begin{aligned}
\|f\|^2 &= \langle f, f \rangle \\
&= \langle \langle f, \varphi \rangle \varphi + zLf, \langle f, \varphi \rangle \varphi + zLf \rangle \\
&= |\langle f, \varphi \rangle|^2 + \|zLf\|^2 \\
&\geqslant \|zLf\|^2 \\
&= \|Lf\|^2,
\end{aligned}$$

where the last equality follows from the integral definition (Proposition 2.3.12) of the norm on $H^2(\widehat{\mathbb{C}} \setminus \gamma)$. Thus L is a contraction on $\mathcal{N}$. $\qquad\square$

Our approach to describing the nearly invariant subspaces of $H^2(\widehat{\mathbb{C}} \setminus \gamma)$ is similar to the one in [7] and is essentially based on a formula relating $\|f\|$ to the values of $(f/\varphi)_i$. To do this, let us first note that since L is a contraction (Proposition 3.2.10), the spectral radius formula says that the spectrum of L is contained in $\mathbb{D}^-$. Thus for every $\lambda \in \mathbb{D}$ the operator $(I - \lambda L)^{-1}$ exists and, by verifying the identity

$$Lf = (I - \lambda L) \left(\frac{f - \frac{f}{\varphi}(\lambda)\varphi}{z - \lambda} \right),$$

we have

$$L_\lambda f := (I - \lambda L)^{-1} Lf = \frac{f - \frac{f}{\varphi}(\lambda)\varphi}{z - \lambda}. \tag{3.2.12}$$

Remark 3.2.13. The alert reader might have some concern over the expression

$$\frac{f}{\varphi}(\lambda)$$

in the above formulas when $\lambda \in \varphi^{-1}(\{0\})$. However, notice that the above formulas are certainly valid for $\lambda \in \mathbb{D} \setminus \varphi^{-1}(\{0\})$ and for these λ's we have

$$\frac{f}{\varphi}(\lambda) = \frac{\lambda (L_\lambda f)(0) + f(0)}{\varphi(0)} = \lambda \langle L_\lambda f, \varphi \rangle + \langle f, \varphi \rangle.$$

By spectral theory, the (far) right-hand side of the above expression is an analytic function on $\mathbb{D}$ (since $L_\lambda = (I - \lambda L)^{-1} L$ is an operator-valued analytic function on $\mathbb{D}$). Thus f/φ extends to be analytic on $\mathbb{D}$.

Lemma 3.2.14. *If $f \in \mathcal{N}$ and $|\lambda| < 1$, then*

$$\|f\|^2 = \left| \frac{f}{\varphi}(\lambda) \right|^2 + (1 - |\lambda|^2) \|L_\lambda f\|^2 - 2\Re \langle f, \lambda L_\lambda f \rangle. \tag{3.2.15}$$

In particular,

$$\frac{f_i}{\varphi_i} \in H^2(\mathbb{D}) \quad \text{and} \quad \left\| \frac{f_i}{\varphi_i} \right\|_{H^2(\mathbb{D})} \leqslant \|f\|.$$

Proof. To prove the formula in (3.2.15), let $f \in \mathcal{N}$ and $\lambda \in \mathbb{D}$ and start with the formula

$$L_\lambda f = \frac{f - \frac{f}{\varphi}(\lambda)\varphi}{z - \lambda}.$$

Manipulate this to get

$$z L_\lambda f = f - \frac{f}{\varphi}(\lambda)\varphi + \lambda L_\lambda f.$$

Now use the identity $\|z L_\lambda f\| = \|L_\lambda f\|$ (Proposition 2.3.12) and the previous formula to get

$$\left\| f - \frac{f}{\varphi}(\lambda)\varphi + \lambda L_\lambda f \right\|^2 = \|L_\lambda f\|^2.$$

Write the left-hand side of the previous equation as an inner product, i.e., $\|\cdot\|^2 = \langle \cdot, \cdot \rangle$, and then multiply it out to obtain

$$\|f\|^2 - 2\Re\left(\overline{\frac{f}{\varphi}(\lambda)} \langle f, \varphi \rangle \right) + \left| \frac{f}{\varphi}(\lambda) \right|^2 + 2\Re\left\langle f - \frac{f}{\varphi}(\lambda)\varphi, \lambda L_\lambda f \right\rangle = (1 - |\lambda|^2) \|L_\lambda f\|^2.$$

We know from (3.2.1) that for $\lambda \neq 0$,

$$\langle \varphi, \lambda L_\lambda f \rangle = \overline{\left(\frac{(\lambda L_\lambda f)(0)}{\varphi(0)} \right)} = \overline{\frac{f}{\varphi}(\lambda)} - \overline{\frac{f}{\varphi}(0)}.$$

Combine this with the previous equation to obtain the identity in (3.2.15).

To show that $(f/\varphi)_i \in H^2(\mathbb{D})$ and $\|f_i/\varphi_i\|_{H^2(\mathbb{D})} \leqslant \|f\|$, use the formula in (3.2.15) to get

$$\left| \frac{f}{\varphi}(\lambda) \right|^2 \leqslant \left| \frac{f}{\varphi}(\lambda) \right|^2 + (1 - |\lambda|^2)\|\lambda L_\lambda f\|^2$$

$$= \|f\|^2 + 2\Re\langle f, \lambda L_\lambda f \rangle.$$

This says that the subharmonic function $|f/\varphi|^2$ on $\mathbb{D}$ has a harmonic majorant

$$\lambda \mapsto \|f\|^2 + 2\Re\langle f, \lambda L_\lambda f \rangle$$

whose value at the origin is $\|f\|^2$. Note the use of Remark 3.2.13 in saying that f/φ is analytic and $\lambda \mapsto \Re\langle f, \lambda L_\lambda f \rangle$ is harmonic on $\mathbb{D}$. The inequality $\|f/\varphi\|_{H^2(\mathbb{D})} \leqslant \|f\|$ now follows from the identity

$$\|g\|^2_{H^2(\mathbb{D})} = U_g(0), \quad g \in H^2(\mathbb{D}),$$

where U_g is the least harmonic majorant of g. $\qquad\square$

This next result is the key step in proving Theorem 3.1.2. In what follows,

$$\omega = \omega_{\widehat{\mathbb{C}} \setminus \gamma, \phi_\gamma(0)}$$

will denote harmonic measure for $\widehat{\mathbb{C}} \setminus \gamma$ at $\phi(0)$, where $\phi : \mathbb{D} \to \widehat{\mathbb{C}} \setminus \gamma$ is the conformal map from (2.3.11).

Theorem 3.2.16. *For every $f \in \mathcal{N}$ we have*

$$\|f\|^2 = \left\| \frac{f}{\varphi} \right\|_{H^2(\mathbb{D})}^2 + \int_\gamma \left| f_e - \frac{f_i}{\varphi_i} \varphi_e \right|^2 d\omega.$$

The proof of this theorem requires a technical lemma. Let

$$P_z(\zeta) := \frac{1 - |z|^2}{|\zeta - z|^2}, \quad z \in \mathbb{D}, \quad \zeta \in \mathbb{T},$$

be the Poisson kernel and recall a classical theorem of Fatou [31, p. 4] which says that for each $f \in L^1(m)$,

$$\lim_{r \to 1^-} \int_{\mathbb{T}} f(\zeta) P_{r\xi}(\zeta) dm(\zeta) = f(\xi) \quad \text{a.e. } \xi \in \mathbb{T}. \tag{3.2.17}$$

It is also well known [45, p. 33] that the above limit holds in the L^1 norm. A slight generalization of this is the following.

Lemma 3.2.18. *For each $f, g \in H^2(\mathbb{D})$, there is a sequence $r_n \nearrow 1$ such that*

$$\lim_{n \to \infty} \int_{\mathbb{T}} f(r_n \zeta) \overline{g(r_n \zeta)} P_{r_n \xi}(\zeta) dm(\zeta) = f(\xi) \overline{g(\xi)} \quad \text{a.e. } \xi \in \mathbb{T}.$$

Proof. For $r \in (0,1)$ and $h \in L^1(m)$, define $h_r(\zeta) := h(r\zeta)$ on $\mathbb{T}$. By the Cauchy-Schwarz inequality, $fg \in H^1(\mathbb{D})$ and so [31, p. 21]

$$f_r g_r \to fg \text{ in } L^1(m) \text{ as } r \to 1^-.$$

Furthermore, $g_r \to g$ almost everywhere as $r \to 1^-$. The identity

$$f_r \overline{g_r} - f\overline{g} = (f_r g_r - fg) \frac{\overline{g_r}}{g_r} + fg \left(\frac{\overline{g_r}}{g_r} - \frac{\overline{g}}{g} \right)$$

along with the above observations and the dominated convergence theorem will show that

$$f_r \overline{g_r} \to f\overline{g} \quad \text{in } L^1(m) \text{ as } r \to 1^-. \tag{3.2.19}$$

For $r \in (0,1)$ and $\xi \in \mathbb{T}$, let

$$F(r, \xi) := \int_{\mathbb{T}} \left(f_r(\zeta) \overline{g_r(\zeta)} - f(\zeta) \overline{g(\zeta)} \right) P_{r\xi}(\zeta) dm(\zeta).$$

By reversing the order of integration and using (3.2.19), along with the identities

$$P_{r\xi}(\zeta) = P_{r\zeta}(\xi), \quad \int_{\mathbb{T}} P_{r\zeta}(\xi)dm(\xi) = 1,$$

we see that $F(r,\cdot) \to 0$ in $L^1(m)$ as $r \to 1^-$. Thus, for some sequence $r_n \uparrow 1$, $F(r_n,\xi) \to 0$ for almost every $\xi \in \mathbb{T}$. Combine all this with Fatou's theorem (3.2.17) to complete the proof. □

Proof of Theorem 3.2.16. Start with the identity

$$\|f\|^2 = \left|\frac{f}{\varphi}(\lambda)\right|^2 + (1 - |\lambda|^2)\|L_\lambda f\|^2 - 2\Re\langle f, \lambda L_\lambda f\rangle$$

from (3.2.15) and let $\lambda = r\zeta$, where $r \in (0,1)$ and $\zeta \in \mathbb{T}$. Now integrate with respect to (normalized) Lebesgue measure $dm = d\theta/2\pi$ on $\mathbb{T}$, and use the mean value theorem for harmonic functions on the last term to get

$$\|f\|^2 = \int_{\mathbb{T}} \left|\frac{f}{\varphi}(r\zeta)\right|^2 dm(\zeta) + (1 - r^2)\int_{\mathbb{T}} \|L_{r\zeta}f\|^2 dm(\zeta).$$

Use the fact from Lemma 3.2.14 that $(f/\varphi)_i \in H^2(\mathbb{D})$ and take a limit as $r \to 1^-$ to obtain

$$\|f\|^2 = \left\|\frac{f}{\varphi}\right\|^2_{H^2(\mathbb{D})} + \lim_{r\to 1^-}(1 - r^2)\int_{\mathbb{T}} \|L_{r\zeta}f\|^2 dm(\zeta). \tag{3.2.20}$$

Note that the second term on the right-hand side of the above equation, without including the limit, can be written (by reversing the order of integration and using the identity $P_{r\xi}(\zeta) = P_{r\zeta}(\xi)$) as

$$(1 - r^2)\int_{\mathbb{T}} \|L_{r\zeta}f\|^2 dm(\zeta)$$

$$= \int_\gamma \int_{\mathbb{T}} \left(\left|f_i(\xi) - \frac{f}{\varphi}(r\zeta)\varphi_i(\xi)\right|^2 + \left|f_e(\xi) - \frac{f}{\varphi}(r\zeta)\varphi_e(\xi)\right|^2\right) P_{r\xi}(\zeta)dm(\zeta)d\omega(\xi).$$

Use the fact that $(f/\varphi)_i \in H^2(\mathbb{D})$ (Lemma 3.2.14) along with Lemma 3.2.18 and Fatou's lemma, to obtain

$$\|f\|^2 \geqslant \left\|\frac{f_i}{\varphi_i}\right\|^2_{H^2(\mathbb{D})} + \int_\gamma \left|f_e - \frac{f_i}{\varphi_i}\varphi_e\right|^2 d\omega.$$

The previous inequality says that the map

$$f \mapsto \left(\frac{f_i}{\varphi_i}, f_e - \frac{f_i}{\varphi_i}\varphi_e\right)$$

is a contraction from $\mathcal{N}$ to $H^2(\mathbb{D}) \oplus L^2(\gamma,\omega)$. In order to show this map is isometric, it suffices to show it for a dense subset of $\mathcal{N}$. In particular, we will show in a moment that this map is isometric on the subset of $f \in \mathcal{N}$ for which $(f/\varphi)_i \in H^\infty(\mathbb{D})$. However by Corollary 3.2.9, this set contains all the finite linear combinations of reproducing kernels for $\mathcal{N}$ – which are dense in $\mathcal{N}$. The fact that this map is isometric on such f follows from (3.2.20), the equality following (3.2.20), Lemma 3.2.18, and the dominated convergence theorem. $\qquad\square$

3.3   The operator J

Let S be the unilateral shift on $H^2 := H^2(\mathbb{D})$, i.e.,

$$(Sh)(z) = zh(z)$$

and T be the operator of multiplication by ζ on $L^2(\gamma, \omega)$, i.e.,

$$(Tk)(\zeta) = \zeta k(\zeta).$$

Observe that

$$(S^*h)(z) = \frac{h(z) - h(0)}{z} \quad \text{and} \quad (T^*k)(\zeta) = \overline{\zeta} k(\zeta).$$

Define

$$M_\zeta : H^2 \oplus L^2(\gamma, \omega) \to H^2 \oplus L^2(\gamma, \omega), \quad M_\zeta := S \oplus T.$$

Corollary 3.3.1. *The map $J : \mathcal{N} \to H^2 \oplus L^2(\gamma, \omega)$ defined by*

$$Jf = \left(\frac{f_i}{\varphi_i}, f_e - \frac{f_i}{\varphi_i}\varphi_e \right)$$

is an isometry and $(J\mathcal{N})^\perp$ is an invariant subspace for M_ζ.

Proof. The first part is just a reformulation of Theorem 3.2.16. To see the second part, note that

$$JLf = \left(S^* \frac{f_i}{\varphi_i}, T^* \left(f_e - \frac{f_i}{\varphi_i}\varphi_e \right) \right) = M_\zeta^* Jf, \quad f \in \mathcal{N}.$$

Since $L\mathcal{N} \subset \mathcal{N}$, we see that $J\mathcal{N}$ is M_ζ^*-invariant and so $(J\mathcal{N})^\perp$ is M_ζ-invariant. $\qquad\square$

The proof of this next lemma needs a fact about Cauchy transforms, known in various circles as 'Fatou's jump theorem'. If $f \in L^1(m)$, the *Cauchy transform*

$$(Cf)(z) := \int_{\mathbb{T}} \frac{f(\zeta)}{\zeta - z} dm(\zeta) \tag{3.3.2}$$

is an analytic function on $\widehat{\mathbb{C}} \setminus \mathbb{T}$. Classical function theory [31, p. 39] says that $(Cf)_i \in H^p(\mathbb{D})$ and $(Cf)_e \in H^p(\mathbb{D}_e)$ for all $0 < p < 1$ and so Cf has finite non-tangential limits $(Cf)_i(\zeta)$ and $(Cf)_e(\zeta)$ for almost every $\zeta \in \mathbb{T}$.

Theorem 3.3.3 (Fatou's jump theorem). *For almost every* $\zeta \in \mathbb{T}$,

$$(Cf)_i(\zeta) - (Cf)_e(\zeta) = \overline{\zeta} f(\zeta).$$

Proof. A computation shows that for $\zeta \in \mathbb{T}$ and $r \in (0,1)$

$$(Cf)(r\zeta) - (Cf)(\zeta/r) = \int_{\mathbb{T}} P_{r\zeta}(\xi)\overline{\xi} f(\xi)dm(\xi),$$

where $P_{r\zeta}(\xi)$ is the usual Poisson kernel. The result now follows from Fatou's theorem (see (3.2.17)). $\qquad\square$

Remark 3.3.4. Fatou's jump theorem can be greatly generalized in many directions [15, 57].

Recall that $\omega \ll m$ and so define

$$w = \frac{d\omega}{dm}.$$

Lemma 3.3.5. *If* $h = (h_1, h_2) \in (J\mathcal{N})^{\perp}$ *and* $f \in \mathcal{N}$, *then*

$$(h_1 - w\chi_\gamma h_2 \overline{\varphi_e})\left(f_e - \frac{f_i}{\varphi_i}\varphi_e\right) = 0 \quad a.e. \ on \ \gamma.$$

Proof. Let $Z = Z(\mathcal{N}) \cup \varphi^{-1}(\{0\})$. If $\lambda \in (\widehat{\mathbb{C}} \setminus \gamma) \setminus Z$ and $f \in \mathcal{N}$, let

$$R_\lambda f := \frac{f - \frac{f}{\varphi}(\lambda)\varphi}{z - \lambda}$$

and note from (3.2.12) that when $|\lambda| < 1$, $R_\lambda f = L_\lambda f$. Also notice, since $\mathcal{N}$ is nearly invariant (and λ is not in the common zero set for $\mathcal{N}$), that $R_\lambda f \in \mathcal{N}$ (Proposition 3.1.1). Using the definition of the operator J, one can show that

$$JR_\lambda f = \left(\frac{\frac{f_i}{\varphi_i} - \frac{f}{\varphi}(\lambda)}{z - \lambda}, \frac{1}{z - \lambda}\left(f_e - \frac{f_i}{\varphi_i}\varphi_e\right)\right).$$

For $h = (h_1, h_2) \in (J\mathcal{N})^{\perp}$, we use the fact that $JR_\lambda f \perp h$ for all $\lambda \in (\widehat{\mathbb{C}} \setminus \gamma) \setminus Z$, to see that

$$\langle JL_{r\zeta}f, h\rangle_{H^2 \oplus L^2(\gamma,\omega)} - \langle JR_{\zeta/r}f, h\rangle_{H^2 \oplus L^2(\gamma,\omega)} = 0$$

whenever $r \in (0,1)$ and $\frac{1}{r}\zeta \notin Z$. The left-hand side of the above can be written in Cauchy transform notation (3.3.2) as

$$C\left(\frac{f_i}{\varphi_i}\overline{h_1}\right)(r\zeta) - \frac{f}{\varphi}(r\zeta)C(\overline{h_1})(r\zeta) + C\left(\left(f_e - \frac{f_i}{\varphi_i}\varphi_e\right)w\overline{h_2}\chi_\gamma\right)(r\zeta) \qquad (3.3.6)$$

$$-C\left(\frac{f_i}{\varphi_i}\overline{h_1}\right)(\zeta/r) + \frac{f}{\varphi}(\zeta/r)C(\overline{h_1})(\zeta/r) - C\left(\left(f_e - \frac{f_i}{\varphi_i}\varphi_e\right)w\overline{h_2}\chi_\gamma\right)(\zeta/r).$$

As $r \to 1^-$, Fatou's jump theorem (Theorem 3.3.3) says that the sum of the first and fourth terms in (3.3.6) approaches

$$\overline{\zeta} \frac{f_i}{\varphi_i} \overline{h_1} \quad \text{a.e.}$$

As $r \to 1^-$, another application of Fatou's jump theorem says that the sum of the third and sixth terms approaches

$$\overline{\zeta} \left(f_e - \frac{f_i}{\varphi_i} \varphi_e \right) w \overline{h_2} \chi_\gamma \quad \text{a.e.}$$

The second term is

$$\frac{f}{\varphi}(r\zeta) C(\overline{h_1})(r\zeta) = \frac{f}{\varphi}(r\zeta) \int_{\mathbb{T}} \frac{\overline{\xi h_1(\xi)}}{1 - \overline{\xi} r\zeta} dm(\xi)$$

$$= \frac{f}{\varphi}(r\zeta) \sum_{n=0}^{\infty} r^n \zeta^n \int_{\mathbb{T}} \overline{\xi}^{n+1} \overline{h_1(\xi)} dm(\xi)$$

$$= 0$$

since $h_1 \in H^2(\mathbb{D})$. Using a similar type of power series computation, the fifth term is

$$\frac{f}{\varphi}(\zeta/r) C(\overline{h_1})(\zeta/r) = \frac{f}{\varphi}(\zeta/r) \left(-\frac{r}{\zeta} \right) \overline{h_1(r\zeta)}$$

which, as $r \to 1^-$, approaches

$$-\overline{\zeta} \frac{f_e}{\varphi_e} \overline{h_1} \quad \text{a.e.}$$

Taking limits as $r \to 1^-$ in (3.3.6) and using the above computations, we get

$$0 = \overline{\zeta} \frac{f_i}{\varphi_i} \overline{h_1} - \overline{\zeta} \frac{f_e}{\varphi_e} \overline{h_1} + \overline{\zeta} \left(f_e - \frac{f_i}{\varphi_i} \varphi_e \right) w \overline{h_2} \chi_\gamma$$

$$= -\overline{\zeta} \left(\frac{\overline{h_1}}{\varphi_e} - w \overline{h_2} \chi_\gamma \right) \left(f_e - \frac{f_i}{\varphi_i} \varphi_e \right)$$

$$= -\frac{\overline{\zeta}}{\varphi_e} \left(\overline{h_1} - w \varphi_e \overline{h_2} \chi_\gamma \right) \left(f_e - \frac{f_i}{\varphi_i} \varphi_e \right)$$

almost everywhere and so

$$\left(\overline{h_1} - w \varphi_e \overline{h_2} \chi_\gamma \right) \left(f_e - \frac{f_i}{\varphi_i} \varphi_e \right) = 0$$

almost everywhere. Using the trivial fact that $ab = 0 \Leftrightarrow \overline{a}b = 0$ the result now follows. $\square$

3.4   The Wold decomposition

Our next step is to analyze the von Neumann-Wold decomposition of M_ζ on the Hilbert space

$$\mathcal{H} := (J\mathcal{N})^\perp.$$

If A is an isometry on a Hilbert space $\mathcal{K}$, then

$$\mathcal{K} = \mathcal{K}' \oplus \mathcal{K}'',$$

where

$$\mathcal{K}' = \bigcap_{n \geqslant 0} A^n \mathcal{K},$$

is reducing for A, i.e., $A\mathcal{K}' \subset \mathcal{K}'$ and $A^* \mathcal{K}' \subset \mathcal{K}'$. Moreover,

$$\mathcal{K}'' = \bigoplus_{n=0}^{\infty} \mathcal{K}_n, \tag{3.4.1}$$

where $\mathcal{K}_n = A^n \mathcal{K} \ominus A^{n+1} \mathcal{K}$ and A maps $\mathcal{K}_n$ isometrically onto $\mathcal{K}_{n+1}$. We say that $A|\mathcal{K}''$ is a *unilateral shift*. Since the dimensions of $\mathcal{K}_n$ must all be the same, we call the common dimension the *multiplicity* of $A|\mathcal{K}''$. In fact, the multiplicity of $A|\mathcal{K}''$ is also equal to the dimension of $\mathcal{K}'' \ominus (A\mathcal{K}'')$. See [24, p. 112] for the details on all of this.

Let us apply the Wold decomposition to the Hilbert space $\mathcal{H} = (J\mathcal{N})^\perp$ and the isometry $M_\zeta|\mathcal{H}$ (Corollary 3.3.1). Here we have

$$\mathcal{H} = \mathcal{H}_0 \oplus \mathcal{H}_1$$

and

$$\mathcal{H}_0 = \bigcap_{n \geqslant 0} M_\zeta^n \mathcal{H}$$

is a reducing subspace for M_ζ, and $M_\zeta|\mathcal{H}_1$ is a unilateral shift.

Proposition 3.4.2. *There is a measurable set $E = E(\mathcal{N}) \subset \gamma$ such that*

$$\mathcal{H}_0 = \{0\} \oplus \chi_E L^2(\gamma, \omega).$$

Moreover, if $F \subset \gamma$ and

$$J\mathcal{N} \subset H^2 \oplus \chi_{F^c} L^2(\gamma, \omega),$$

then $m(F \setminus E) = 0$.

Proof. To see the first part, recall that $M_\zeta = S \oplus T$, where

$$(Sf)(z) = zf(z), \quad f \in H^2 := H^2(\mathbb{D})$$

and

$$(Tg)(\zeta) = \zeta g(\zeta), \quad g \in L^2(\gamma, \omega).$$

Since

$$\bigcap_{n \geqslant 0} S^n H^2 = \{0\},$$

we see that

$$\mathcal{H}_0 = \bigcap_{n \geqslant 0} (S \oplus T)^n \mathcal{H} = \{0\} \oplus Y,$$

where Y is a T-invariant subspace of $L^2(\gamma, \omega)$. Furthermore, since $\gamma \neq \mathbb{T}$, we know that both $\zeta Y \subset Y$ and $\overline{\zeta} Y \subset Y$ hold (use Lavrentiev's theorem [22, p. 232] to approximate the function $v(\zeta) = \overline{\zeta}$ uniformly on γ by a sequence of analytic polynomials). Now apply a classical theorem of Wiener [43, p. 7] to get

$$Y = \chi_E L^2(\gamma, \omega)$$

for some measurable subset $E \subset \mathbb{T}$.

To see the second part of the proposition, observe that a routine argument shows that

$$(H^2 \oplus \chi_{F^c} L^2(\gamma, \omega))^\perp = \{0\} \oplus \chi_F L^2(\gamma, \omega).$$

Thus, if we assume that

$$J\mathcal{N} \subset H^2 \oplus \chi_{F^c} L^2(\gamma, \omega),$$

we have

$$\mathcal{H} = (J\mathcal{N})^\perp \supset \{0\} \oplus \chi_F L^2(\gamma, \omega)$$

and so

$$\mathcal{H}_0 = \bigcap_{n \geqslant 0} M_\zeta^n \mathcal{H} \supset \{0\} \oplus \chi_F L^2(\gamma, \omega).$$

But from the first part of the theorem,

$$\mathcal{H}_0 = \{0\} \oplus \chi_E L^2(\gamma, \omega)$$

and the result follows.   □

Lemma 3.4.3. *Let* $\mathcal{H}_0 = \{0\} \oplus \chi_E L^2(\gamma, \omega)$ *as in Proposition* 3.4.2. *Then any* $\phi = (a, b) \in \mathcal{H}_1 \ominus M_\zeta \mathcal{H}_1$ *of unit norm, must satisfy*

$$a \in H^\infty(\mathbb{D}); \tag{3.4.4}$$

$$|a|^2 + |b|^2 w = 1 \quad a.e.\ on\ \mathbb{T}^4; \tag{3.4.5}$$

$$b = 0 \quad a.e.\ on\ E; \tag{3.4.6}$$

$$a = b w \overline{\varphi_e} \quad a.e\ on\ \gamma \setminus E. \tag{3.4.7}$$

---

[4]We extend the domain of b to be all of $\mathbb{T}$ by defining it to be zero on $\mathbb{T} \setminus \gamma$.

Proof. Let $\phi = (a,b) \in \mathcal{H}_1 \ominus M_\zeta \mathcal{H}_1$ of unit norm. Extend the domain of b to include the entire unit circle by setting it to be zero on $\mathbb{T} \setminus \gamma$. From here it follows that for each $n \in \mathbb{N}$,

$$0 = \left\langle (a,b), M_\zeta^n(a,b) \right\rangle_{H^2 \oplus L^2(\gamma,\omega)} = \int_{\mathbb{T}} (|a|^2 + |b|^2 w) \overline{\zeta}^n dm(\zeta).$$

Taking complex conjugates, we see that all of the non-zero Fourier coefficients of $|a|^2 + |b|^2 w$ are equal to zero and so this function is a (non-negative) constant. Since ϕ has unit norm, this constant must be equal to 1. This proves (3.4.5) as well as the fact that $a \in H^\infty(\mathbb{D})$ (since it is an H^2 function with bounded boundary values).

Since $\phi = (a,b) \in \mathcal{H}_1 = (\mathcal{H}_0)^\perp = (\{0\} \oplus \chi_E L^2(\gamma,\omega))^\perp$ (Proposition 3.4.2), we know that

$$\int b \chi_E \overline{g} w \, dm = 0 \quad \text{for all } g \in L^2(\gamma,\omega).$$

From basic measure theory, it follows that $b = 0$ almost everywhere on E, proving (3.4.6).

By Lemma 3.3.5 we know that for every $f \in \mathcal{N}$,

$$(a - bw\overline{\varphi_e}) \left(f_e - \frac{f_i}{\varphi_i} \varphi_e \right) = 0 \quad \text{a.e. on } \gamma.$$

If $F \subset \gamma \setminus E$ has positive measure, the second statement in Proposition 3.4.2 says that there is an $f \in \mathcal{N}$ so that

$$f_e - \frac{f_i}{\varphi_i} \varphi_e$$

is non-zero on some subset F' of F of positive measure. It follows that $a - bw\overline{\varphi_e} = 0$ almost everywhere on F' and hence zero almost everywhere on $\gamma \setminus E$. This proves (3.4.7). $\qquad\square$

Proposition 3.4.8. *$M_\zeta | \mathcal{H}_1$ is a unilateral shift of multiplicity* 1.

Proof. From our earlier discussion of multiplicities of shifts, we need to show that $\mathcal{H}_1 \ominus M_\zeta \mathcal{H}_1$ is one-dimensional. Let $\phi_j = (a_j, b_j), j = 1, 2$, belong to $\mathcal{H}_1 \ominus M_\zeta \mathcal{H}_1$ with $\phi_1 \perp \phi_2$. Recall from Lemma 3.4.3 that $a_j, b_j, j = 1, 2$, satisfy the properties (3.4.6) and (3.4.7). Since $\phi_j \perp M_\zeta^n \mathcal{H}_1$ for all $n \in \mathbb{N}$ and $\phi_1 \perp \phi_2$ we get

$$\left\langle \phi_1, M_\zeta^n \phi_2 \right\rangle_{H^2 \oplus L^2(\gamma,\omega)} = 0 \quad \forall n \in \mathbb{N}_0,$$

$$\left\langle \phi_2, M_\zeta^n \phi_1 \right\rangle_{H^2 \oplus L^2(\gamma,\omega)} = 0 \quad \forall n \in \mathbb{N}_0.$$

The first equation says that

$$\int_{\mathbb{T}} (a_1 \overline{a_2} \overline{\zeta}^n + b_1 \overline{b_2} \overline{\zeta}^n w) dm = 0 \quad \forall n \in \mathbb{N}_0.$$

The second equation says that

$$\int_{\mathbb{T}} (a_2 \overline{a_1} \overline{\zeta}^n + b_2 \overline{b_1} \overline{\zeta}^n w) dm = 0 \quad \forall n \in \mathbb{N}_0.$$

The complex conjugate of the above equation, together with one just before it, say that all of the Fourier coefficients of the function $a_1\overline{a_2} + b_1\overline{b_2}w$ vanish. Thus

$$a_1\overline{a_2} + b_1\overline{b_2}w = 0 \quad \text{a.e. on } \mathbb{T}.$$

This means that either a_1 or a_2 must vanish identically (since we are assuming, as in Lemma 3.4.3, that we have extended the domains of b_1, b_2 to all of $\mathbb{T}$ by making them zero off γ) and so by (3.4.6) and (3.4.7), either ϕ_1 or ϕ_2 must vanish identically. $\qquad\square$

The operator $M_\zeta | \mathcal{H}_1$ is a shift of multiplicity 1 (Proposition 3.4.8) and so if $\phi \in \mathcal{H}_1 \ominus M_\zeta \mathcal{H}_1$ and has norm 1, then $\phi = (a, b)$ satisfies the properties of Lemma 3.4.3 as well as, via the consequences of the Wold decomposition in (3.4.1),

$$\mathcal{H}_1 = \bigvee_{n=0}^{\infty} M_\zeta^n \phi.^5 \tag{3.4.9}$$

The function a satisfies one more property. But first we need a few definitions. Recall that $\mathbb{D}_e := \widehat{\mathbb{C}} \setminus \mathbb{D}^-$, and define

$$H_0^2(\mathbb{D}_e) := \{ f \in H^2(\mathbb{D}_e) : f(\infty) = 0 \},$$

$$N_0^+(\mathbb{D}_e) := \{ f \in N^+(\mathbb{D}_e) : f(\infty) = 0 \}.$$

The space $N^+(\mathbb{D}_e)$, the Smirnov class, was defined in (2.4.4).

Lemma 3.4.10. *If Θ_e denotes the greatest common $\mathbb{D}_e$-inner factor of $\{ f_e : f \in \mathcal{N} \}$, the function*

$$z \mapsto \frac{\overline{a}(1/\overline{z})\Theta_e(z)}{\varphi_e(z)}, \quad z \in \mathbb{D}_e,$$

belongs to $N_0^+(\mathbb{D}_e)$.

Proof. Using the fact that $\mathcal{H}_1 \subset (J\mathcal{N})^\perp$ and (3.4.9), we see, for all $f \in \mathcal{N}$ that

$$\left\langle \frac{f}{\varphi}, az^n \right\rangle_{H^2} + \left\langle \left(f_e - \frac{f_i}{\varphi_i}\varphi_e \right), b\zeta^n \right\rangle_{L^2(\gamma, \omega)} = 0 \quad \forall n \in \mathbb{N}_0.$$

Write the above identity out as an integral and use the F. and M. Riesz theorem [6] [31, p. 41] to see that

$$\frac{f_i}{\varphi_i}\overline{a} + \left(f_e - \frac{f_i}{\varphi_i}\varphi_e \right) w\overline{b}\chi_\gamma \in \overline{H_0^2}.$$

More precisely, the above function is equal to $\overline{h}$ almost everywhere on $\mathbb{T}$, where $h \in H_0^2$.

---

[5] Here we are using the general fact that if V is a shift of multiplicity n on a Hilbert space $\mathcal{X}$ and $\{ g : 1 \leqslant j \leqslant n \}$ is a basis for $\mathcal{X} \ominus (V\mathcal{X})$, then $\{ V^k e_j : k \in \mathbb{N}_0, 1 \leqslant j \leqslant n \}$ is a basis for $\mathcal{X}$. See [24, Prop. 23.10].

[6] F. and M. Riesz theorem: If μ is a finite complex measure on the unit circle $\mathbb{T}$ whose Fourier coefficients $\hat{\mu}(n)$ vanish for all $n \in \mathbb{N}$, then $d\mu = \overline{f}\,dm$ for some $f \in H^1$.

On $\mathbb{T} \setminus \gamma$,

$$\overline{h} = \frac{f_i}{\varphi_i}\overline{a} = \frac{f_e}{\varphi_e}\overline{a}$$

and the function

$$z \mapsto \frac{f_e(z)}{\varphi_e(z)}\overline{a}(1/\overline{z}) \tag{3.4.11}$$

belongs to the Nevanlinna class of $\mathbb{D}_e$. Since $\overline{h}(1/\overline{z}) \in H_0^2(\mathbb{D}_e)$, and has the same non-tangential boundary values as the function in (3.4.11) on $\mathbb{T} \setminus \gamma$, these two functions must be identical (Privalov's uniqueness theorem - see [51, p. 62]).

Since the function in (3.4.11) belongs to $H_0^2(\mathbb{D}_e)$ for all $f \in \mathcal{N}$ and Θ_e is the greatest common $\mathbb{D}_e$-inner factor of $\{f_e : f \in \mathcal{N}\}$, the result now follows. $\qquad\square$

The final technical lemma needed to prove Theorem 3.1.2 is the following.

Lemma 3.4.12. *Suppose $\mathcal{N}$ is a subspace of $H^2(\widehat{\mathbb{C}} \setminus \gamma)$ with greatest common $\widehat{\mathbb{C}} \setminus \gamma$-inner divisor one. Then the greatest common $\mathbb{D}$-inner divisor of $\mathcal{N}|\mathbb{D}$ is equal to 1.*

Proof. Let

$$\mathcal{N}_1 = \mathrm{clos}_{H^2(\widehat{\mathbb{C}} \setminus \gamma)}(H^\infty(\widehat{\mathbb{C}} \setminus \gamma)\mathcal{N})$$

and note that $H^\infty(\widehat{\mathbb{C}} \setminus \gamma)\mathcal{N}_1 \subset \mathcal{N}_1$. Using the same proof as Proposition 6.1.2 below, we see, since the greatest common $\widehat{\mathbb{C}} \setminus \gamma$-inner divisor of $\mathcal{N}$ is 1, that $\mathcal{N}_1 = H^2(\widehat{\mathbb{C}} \setminus \gamma)$. By the definition of $\mathcal{N}_1$, there are $\phi_n \in H^\infty(\widehat{\mathbb{C}} \setminus \gamma)$ and $f_n \in \mathcal{N}$ so that

$$\phi_n f_n \to \chi_{\widehat{\mathbb{C}} \setminus \gamma} \quad \text{in } H^2(\widehat{\mathbb{C}} \setminus \gamma) \text{ as } n \to \infty.$$

It follows from (2.1.4) that

$$(\phi_n f_n)_i \to \chi_{\mathbb{D}} \quad \text{in } H^2(\mathbb{D}) \text{ as } n \to \infty.$$

Suppose that $\mathcal{N}|\mathbb{D}$ has a non-constant greatest common $\mathbb{D}$-inner divisor ϑ. Then the previous equation says that

$$\vartheta\left(\frac{(\phi_n f_n)_i}{\vartheta} - \overline{\vartheta}\right) \to 0 \quad \text{in the norm of } L^2(m)$$

and consequently

$$\frac{(\phi_n f_n)_i}{\vartheta} \to \overline{\vartheta} \quad \text{in the norm of } L^2(m).$$

But since $(\phi_n f_n)_i/\vartheta \in H^2(\mathbb{D})$ for all n, then $\overline{\vartheta} \in H^2(\mathbb{D})$, which is not the case. Thus, by contradiction, $\mathcal{N}|\mathbb{D}$ has greatest common $\mathbb{D}$-inner divisor 1. $\qquad\square$

3.5   Proof of the main theorem

We are now ready for the proof of Theorem 3.1.2.

Proof of Theorem 3.1.2. Before getting to the crux of the proof, let us set up a few things and remind the reader what we have already shown.

Without loss of generality, we can assume that $\Theta_{\mathcal{N}} \equiv 1$. If this is not the case, apply the argument below to $\mathcal{N}/\Theta_{\mathcal{N}}$. Lemma 3.4.12 says that the greatest common $\mathbb{D}$-inner factor of $\{f_i : f \in \mathcal{N}\}$ is 1 as is the greatest common $\mathbb{D}_e$-inner factor of $\{f_e : f \in \mathcal{N}\}$.

Let $E \subset \gamma$ be the measurable set from Proposition 3.4.2 and let F be the $\mathbb{D}$-outer function which satisfies

$$|F|^2 = \frac{|\varphi_i|^2}{1 + w\chi_\gamma|\varphi_e|^2} \quad \text{a.e. on } \mathbb{T}.^{[7]} \tag{3.5.1}$$

Let

$$\rho := \frac{\varphi_i}{\varphi_e} \quad \text{a.e. on } \gamma. \tag{3.5.2}$$

Lemma 3.2.14 says that $(f/\varphi)_i \in H^2(\mathbb{D})$ whenever $f \in \mathcal{N}$. Thus, since we are assuming that $\Theta_{\mathcal{N}} \equiv 1$ (note the discussion in the second paragraph of the proof), it must be the case that

$$\varphi_i \text{ is } \mathbb{D}\text{-outer.} \tag{3.5.3}$$

If a is the $H^\infty(\mathbb{D})$ function from Lemma 3.4.3, we know, again since $\Theta_{\mathcal{N}} \equiv 1$ (and the discussion in the second paragraph of the proof), from Lemma 3.4.10 that

$$\frac{\overline{a}(1/\overline{z})}{\varphi_e(z)} \in N_0^+(\mathbb{D}_e). \tag{3.5.4}$$

Our final reminder is that

$$J\mathcal{N} = \left\{ \left(\frac{f_i}{\varphi_i}, f_e - \frac{f_i}{\varphi_i}\varphi_e \right) : f \in \mathcal{N} \right\} \subset H^2 \oplus L^2(\gamma, \omega)$$

and, from the proof of Proposition 3.4.2 and Lemma 3.2.14,

$$J\mathcal{N} \subset H^2 \oplus \chi_{E^c} L^2(\gamma, \omega),$$

that is to say, for all $f \in \mathcal{N}$ we have

$$\frac{f_i}{\varphi_i} \in H^2, \tag{3.5.5}$$

---

[7]Observe that

$$\log \frac{|\varphi_i|^2}{1 + w\chi_\gamma|\varphi_e|^2} = \log|\varphi_i|^2 - \log(1 + w\chi_\gamma|\varphi_e|^2) \quad a.e.$$

The first summand is integrable on $\mathbb{T}$. Use the fact that $\varphi \in H^2(\mathbb{D}) \setminus \{0\}$ (2.1.4) and (2.4.2) to see this. The second is integrable since $|\varphi_e|^2 w$ is integrable. Thus by standard Nevanlinna theory [31, Ch. 2], such a $\mathbb{D}$-outer function F satisfying (3.5.1) exists.

$$f_e - \frac{f_i}{\varphi_i} \varphi_e \in L^2(\gamma, \omega), \tag{3.5.6}$$

and

$$f_e - \frac{f_i}{\varphi_i} \varphi_e = 0 \quad \text{a.e. on } E. \tag{3.5.7}$$

We are now at the crux of the proof. Let

$$\mathcal{N}_1 := \left\{ f \in H^2(\widehat{\mathbb{C}} \setminus \gamma) : \frac{f_i}{F} \in H^2(\mathbb{D}), f_i = \rho f_e \text{ a.e. on } E \right\}.$$

To show

$$\mathcal{N} = \mathcal{N}_1,$$

we begin with the following claim.

Claim 1: $\mathcal{N} \subset \mathcal{N}_1$.

To prove this claim, we suppose that $f \in \mathcal{N}$ and notice that F is $\mathbb{D}$-outer and so $f_i/F \in N^+(\mathbb{D})$. We now prove that $f_i/F \in H^2(\mathbb{D})$ by showing it has $L^2(m)$ boundary values. Indeed, by the definition of the $\mathbb{D}$-outer function F in (3.5.1),

$$\int_{\mathbb{T}} \left| \frac{f_i}{F} \right|^2 dm = \int_{\mathbb{T}} \left| \frac{f_i}{\varphi_i} \right|^2 (1 + w\chi_\gamma |\varphi_e|^2) \, dm$$

$$= \int_{\mathbb{T}} \left| \frac{f_i}{\varphi_i} \right|^2 dm + \int_\gamma \left| \frac{f_i}{\varphi_i} \varphi_e \right|^2 w \, dm.$$

From (3.5.5) and (3.5.6), both of these integrals converge. Thus f_i/F belongs to $H^2(\mathbb{D})$.

To finish the proof of Claim 1, we can use (3.5.7), to see

$$\frac{f_i}{f_e} = \frac{\varphi_i}{\varphi_e} = \rho \quad \text{a.e. on } E.$$

Thus $f \in \mathcal{N}_1$, which proves Claim 1.

Claim 2: $\mathcal{N}_1 \subset \mathcal{N}$.

Let $g \in \mathcal{N}_1$ and define

$$x_g := \left(\frac{g_i}{\varphi_i}, g_e - \frac{g_i}{\varphi_i} \varphi_e \right).$$

We first want to show that $x_g \in H^2 \oplus L^2(\gamma, \omega)$. By (3.5.3),

$$\frac{g_i}{\varphi_i} \in N^+(\mathbb{D}).$$

To show this function belongs to $H^2(\mathbb{D})$, we need to show it has $L^2(m)$ boundary values. Indeed, from (3.5.1),

$$\int_{\mathbb{T}} \left| \frac{g_i}{\varphi_i} \right|^2 dm = \int_{\mathbb{T}} \left| \frac{g_i}{F} \right|^2 \frac{1}{1 + w\chi_\gamma |\varphi_e|^2} \, dm \leqslant \int_{\mathbb{T}} \left| \frac{g_i}{F} \right|^2 dm.$$

By the definition of $\mathcal{N}_1$, we know that $g_i/F \in H^2(\mathbb{D})$. Thus it follows that

$$\frac{g_i}{\varphi_i} \in H^2(\mathbb{D}). \tag{3.5.8}$$

Furthermore, to show

$$g_e - \frac{g_i}{\varphi_i}\varphi_e \in L^2(\gamma, \omega),$$

notice that the first term belongs to $L^2(\gamma, \omega)$ (from the definition of the $H^2(\widehat{\mathbb{C}} \setminus \gamma)$ norm). For the second term, observe that

$$\int_\gamma \left|\frac{g_i}{\varphi_i}\varphi_e\right|^2 w \, dm = \int_\gamma \left|\frac{g_i}{F}\right|^2 \frac{w|\varphi_e|^2}{1 + w|\varphi_e|^2} \, dm \leqslant \int_\gamma \left|\frac{g_i}{F}\right|^2 dm$$

which is finite since $g_i/F \in H^2(\mathbb{D})$. Thus

$$x_g \in H^2 \oplus L^2(\gamma, \omega).$$

We now want to show

$$x_g \in J\mathcal{N}.$$

We will do this by proving

$$x_g \perp (J\mathcal{N})^\perp.$$

Note here that J is an isometry and so $J\mathcal{N}$ is closed. Thus $x_g \in J\mathcal{N} \Leftrightarrow x_g \perp (J\mathcal{N})^\perp$. Using Proposition 3.4.2 and (3.4.9) we have

$$(J\mathcal{N})^\perp = \mathcal{H}_0 \oplus \mathcal{H}_1 = \left(\{0\} \oplus \chi_E L^2(\gamma, \omega)\right) \oplus \bigvee_{n \geqslant 0} M_\zeta^n \phi.$$

By the definition of $\mathcal{N}_1$, we know that

$$g_e - \frac{g_i}{\varphi_i}\varphi_e = 0 \quad \text{a.e. on } E. \tag{3.5.9}$$

Hence,

$$x_g \perp \{0\} \oplus \chi_E L^2(\gamma, \omega).$$

We are left with showing

$$\langle x_g, M_\zeta^n \phi \rangle_{H^2 \oplus L^2(\gamma,\omega)} = 0 \quad \forall n \in \mathbb{N}_0.$$

If $\phi = (a, b)$ as in (3.4.9) (with the understanding that the domain of b is all of $\mathbb{T}$ be defining it to be zero on $\mathbb{T} \setminus \gamma$), the F. and M. Riesz theorem says that

$$\langle x_g, (z^n a, \zeta^n b) \rangle_{H^2 \oplus L^2(\gamma,\omega)} = 0 \quad \forall n \in \mathbb{N}_0$$

if and only if

$$\frac{g_i}{\varphi_i}\bar{a} + \left(g_e - \frac{g_i}{\varphi_i}\varphi_e\right) w\bar{b} \in \overline{H_0^2}. \tag{3.5.10}$$

By considering three cases: $\mathbb{T} \setminus \gamma$, $\gamma \setminus E$, and E, and using (3.4.6) and (3.4.7), along with the facts that $g_i/\varphi_i = g_e/\varphi_e$ on $\mathbb{T} \setminus \gamma$ and almost everywhere on E, one can show that the function on the left-hand side of (3.5.10) is equal to

$$\frac{g_e}{\varphi_e}\bar{a} \quad \text{a.e. on } \mathbb{T}.$$

But by (3.5.4) we can prove that this function belongs to $H_0^2(\mathbb{D}_e)$ by showing it as $L^2(m)$ boundary values. Indeed,

$$\int_\mathbb{T} \left| \frac{g_e}{\varphi_e}\bar{a} \right|^2 dm = \left(\int_E + \int_{\gamma \setminus E} + \int_{\mathbb{T} \setminus \gamma} \right) \left| \frac{g_e}{\varphi_e}\bar{a} \right|^2 dm. \tag{3.5.11}$$

For the first integral in (3.5.11),

$$\int_E \left| \frac{g_e}{\varphi_e}\bar{a} \right|^2 dm = \int_E \left| \frac{g_i}{\varphi_i}\bar{a} \right|^2 dm.$$

This integral converges since $a \in H^\infty(\mathbb{D})$ and $(g/\varphi)_i \in H^2(\mathbb{D})$ (3.5.8). For the second integral in (3.5.11),

$$\int_{\gamma \setminus E} \left| \frac{g_e}{\varphi_e}\bar{a} \right|^2 dm = \int_{\gamma \setminus E} \left| \frac{g_e}{\varphi_e}\bar{b}\varphi_e w \right|^2 dm \quad \text{(by (3.4.7))}$$

$$= \int_{\gamma \setminus E} |g_e|^2 |b|^2 w w \, dm$$

$$= \int_{\gamma \setminus E} |g_e|^2 (1 - |a|^2) w \, dm \quad \text{(by (3.4.5))}.$$

The above integral converges by the definition of the norm on $H^2(\widehat{\mathbb{C}} \setminus \gamma)$ and the fact that $a \in \text{ball}(H^\infty)$. The third integral in (3.5.11) converges since g_e/φ_e analytically continues to g_i/φ_i across $\mathbb{T} \setminus \gamma$ and so

$$\int_{\mathbb{T} \setminus \gamma} \left| \frac{g_e}{\varphi_e}\bar{a} \right|^2 dm = \int_{\mathbb{T} \setminus \gamma} \left| \frac{g_i}{\varphi_i}\bar{a} \right|^2 dm$$

which converges since $(g/\varphi)_i \in H^2(\mathbb{D})$ (see (3.5.8)) and $a \in H^\infty(\mathbb{D})$.

Thus $x_g \in J\mathcal{N}$ and so $x_g = Jf$ for some $f \in \mathcal{N}$. However, by the definition of the operator J,

$$\frac{g_i}{\varphi_i} = \frac{f_i}{\varphi_i}$$

and so $g = f \in \mathcal{N}$. This proves Claim 2 and hence the main theorem. $\qquad\square$

3.6   Uniqueness of the parameters

Let F be a closed subset of the arc γ. A well-known result of Ahlfors and Beurling [1] or [33] (see p. 6 and p. 29) says that $H^\infty(\widehat{\mathbb{C}} \setminus F)$ contains non-constant functions if and only if $m(F) > 0$.[8] This means, via a version of Morera's theorem [34, p. 95], that if $m(F) > 0$, then there is an $f \in H^\infty(\widehat{\mathbb{C}} \setminus F)$ such that

$$\frac{f_e}{f_i} \neq 1 \quad \text{a.e. on } F', \tag{3.6.1}$$

for some compact $F' \subset F$ with $m(F') > 0$. This observation along with Theorem 3.1.2 yields the following two corollaries.

Corollary 3.6.2. *Let A be a closed subset of γ. Then a nearly invariant subspace $\mathcal{N}$ of the form in (3.1.3) is $H^\infty(\widehat{\mathbb{C}} \setminus A)$-invariant if and only if $m(A \cap E) = 0$.*

Corollary 3.6.3. *Let $\mathcal{N}$ be a nearly invariant subspace of the form in (3.1.3).*

1. *If $\mathcal{N}$ is not $H^\infty(\widehat{\mathbb{C}} \setminus \gamma)$-invariant, then the parameters Θ, E, and ρ are unique in the sense that if Θ_j, E_j, ρ_j, $j = 1, 2$, represent $\mathcal{N}$, then $\Theta_1 = e^{it}\Theta_2$, $m(E_1 \Delta E_2) = 0$, and $\rho_1 \chi_{E_1} = \rho_2 \chi_{E_2}$ almost everywhere.*

2. *If $H^\infty(\widehat{\mathbb{C}} \setminus \gamma)\mathcal{N} \subset \mathcal{N}$, then Θ is unique up to a unimodular constant and $m(E) = 0$.*

Proof. The parameter Θ is the greatest common $\widehat{\mathbb{C}} \setminus \gamma$-inner divisor of the functions in $\mathcal{N}$ and thus is unique up to a unimodular constant.

Suppose there are two subsets E_1, E_2 of γ and functions ρ_1, ρ_2 which represent the same nearly invariant subspace $\mathcal{N}$ as in (3.1.3) and $m(E_1 \setminus E_2) > 0$. Let $\mathcal{N}_{\rho_j, E_j}, j = 1, 2$, denote $\mathcal{N}$ represented by ρ_j, E_j. If A is any closed subset of $E_1 \setminus E_2$ with $m(A) > 0$ and $g \in H^\infty(\widehat{\mathbb{C}} \setminus A)$, the definition of $\mathcal{N}_{\rho_2, E_2}$ says that $g\mathcal{N}_{\rho_2, E_2} \subset \mathcal{N}_{\rho_2, E_2}$. However $\mathcal{N}_{\rho_1, E_1} = \mathcal{N}_{\rho_2, E_2}$ and so $\mathcal{N}_{\rho_1, E_1}$ is $H^\infty(\widehat{\mathbb{C}} \setminus A)$-invariant. The previous corollary says that $m(A \cap E_1) = 0$ which is a contradiction. This says that $m(E_1 \Delta E_2) = 0$. Notice that $H^\infty(\widehat{\mathbb{C}} \setminus \gamma)\mathcal{N} \subset \mathcal{N}$ if and only if $m(E) = 0$.

Suppose $\mathcal{N}$ is not $H^\infty(\widehat{\mathbb{C}} \setminus \gamma)$-invariant. Then $m(E) > 0$. Pick any non-zero function $f \in \mathcal{N}$ and notice that $\rho_1 = \rho_2 = f_i/f_e$ almost everywhere on E. □

---

[8]The general theorem here is, for a compact subset $F \subset \mathbb{C}$, that $H^\infty(\widehat{\mathbb{C}} \setminus F)$ contains non-constant functions if and only if the analytic capacity of F is positive. However, when $F \subset \mathbb{T}$, the analytic capacity is positive if and only if the Lebesgue measure is F is positive.

Chapter 4

Nearly invariant and the backward shift

4.1   The backward shift and pseudocontinuations

For a $\mathbb{D}$-inner function ϑ, form the subspace

$$K_{z\vartheta} := H^2(\mathbb{D}) \cap (z\vartheta H^2(\mathbb{D}))^\perp.$$

Since $z\vartheta H^2(\mathbb{D})$ is an S-invariant subspace of $H^2(\mathbb{D})$, then $K_{z\vartheta}$ will be an S^*-invariant subspace of $H^2(\mathbb{D})$, where

$$S^* f = \frac{f - f(0)}{z}$$

is the backward shift operator. It is also easy to see that $K_{z\vartheta}$ contains the constants. In fact, by Beurling's theorem, every S^*-invariant subspace, which also contains the constants, takes the form $K_{z\vartheta}$ for some $\mathbb{D}$-inner function ϑ. It is well known [16, 26] that functions in $K_{z\vartheta}$ have special 'continuation' properties. Indeed, recall from (3.3.2) that for $h \in L^1(m)$

$$(Ch)(\lambda) := \int_{\mathbb{T}} \frac{h(\zeta)}{\zeta - \lambda} dm(\zeta)$$

denotes the Cauchy transform of h. It is known [16, p. 87] that for any $f \in K_{z\vartheta}$ the meromorphic function

$$\widetilde{f}(\lambda) := \frac{C(f\overline{\zeta}\vartheta)(\lambda)}{C(\overline{\zeta}\vartheta)(\lambda)} \tag{4.1.1}$$

on $\mathbb{D}_e$ is a *pseudocontinuation* of f in that the non-tangential limits of f (from $\mathbb{D}$) and $\widetilde{f}$ (from $\mathbb{D}_e$) are equal almost everywhere on $\mathbb{T}$. Using the Cauchy integral formula and power series, one can prove the identity

$$\widetilde{f}(\lambda) = \frac{1}{\vartheta^*(\lambda)} \sum_{n=1}^{\infty} \frac{1}{\lambda^{n-1}} \widehat{f\overline{\zeta}\vartheta}(-n),$$

where $\widehat{\ }(k)$ denotes the k-th Fourier coefficient and

$$\vartheta^*(\lambda) := \overline{\vartheta\left(\frac{1}{\overline{\lambda}}\right)}, \quad \lambda \in \mathbb{D}_e. \tag{4.1.2}$$

This says that

$$\widetilde{f} \in \frac{1}{\vartheta^*} H^2(\mathbb{D}_e) \quad \forall f \in K_{z\vartheta}. \tag{4.1.3}$$

4.2   A new description of nearly invariant subspaces

The main theorem of this section (Theorem 4.2.5 below) gives an alternate description of the nearly invariant subspaces of $H^2(\widehat{\mathbb{C}} \setminus \gamma)$ in terms of these $K_{z\vartheta}$ spaces. Before getting to this description, we make a remark about norming points.

Remark 4.2.1. If one carefully works through the proof of Theorem 3.1.2 and all the preliminary results that lead up to it, one can see that the result does not depend on the norming point for $H^2(\widehat{\mathbb{C}} \setminus \gamma)$.[1] Up to now, we have been operating under the assumption that the norming point for $H^2(\widehat{\mathbb{C}} \setminus \gamma)$ was $\phi_\gamma(0)$, where $\phi_\gamma = \alpha \circ \phi_G$ (see the appendix for the exact formulas for α and ϕ_G) is a certain conformal map from $\mathbb{D}$ onto $\widehat{\mathbb{C}} \setminus \gamma$. Let us now change the norming point for $H^2(\widehat{\mathbb{C}} \setminus \gamma)$ to be the origin. This yields an equivalent norm on $H^2(\widehat{\mathbb{C}} \setminus \gamma)$ and has the added benefit that

$$\begin{aligned}
\langle f, 1 \rangle_{H^2(\widehat{\mathbb{C}} \setminus \gamma, 0)} &= \langle f \circ \phi_\gamma, 1 \rangle_{H^2(\mathbb{D}, \phi_\gamma^{-1}(0))} \\
&= \int_{\mathbb{T}} (f \circ \phi_\gamma)(\zeta) d\omega_{\mathbb{D}, \phi_\gamma^{-1}(0)} \\
&= \int_{\mathbb{T}} (f \circ \phi_\gamma)(\zeta) \frac{1 - |\phi_\gamma^{-1}(0)|^2}{|\zeta - \phi_\gamma^{-1}(0)|^2} dm(\zeta) \quad \text{(from (2.2.1))} \\
&= f(0)
\end{aligned}$$

and

$$\begin{aligned}
\langle 1, 1 \rangle_{H^2(\widehat{\mathbb{C}} \setminus \gamma, 0)} &= \langle 1, 1 \rangle_{H^2(\mathbb{D}, \phi_\gamma^{-1}(0))} \\
&= \int_{\mathbb{T}} \frac{1 - |\phi_\gamma^{-1}(0)|^2}{|\zeta - \phi_\gamma^{-1}(0)|^2} dm(\zeta) \\
&= 1.
\end{aligned}$$

Thus the constant function 1 is a normalized reproducing kernel for $H^2(\widehat{\mathbb{C}} \setminus \gamma, 0)$. This assumption that the norming point is the origin will be especially important in the proof of Corollary 4.2.20 below.

---

[1] When we need to emphasize the norming point z_0 for $H^2(\Omega)$ we will use the notation $H^2(\Omega, z_0)$

Consider the space $\widetilde{K_{z\vartheta}}$ of meromorphic functions f on $\widehat{\mathbb{C}} \setminus \mathbb{T}$ by $f_i \in K_{z\vartheta}$ and $f_e = \widetilde{f_i}$. Recall the definition of $\widetilde{f}$ from (4.1.1). If $\mathcal{N}$ is a nearly invariant subspace of $H^2(\widehat{\mathbb{C}} \setminus \gamma)$, φ is the extremal function for $\mathcal{N}$, and ϑ_e is the $\mathbb{D}_e$-inner factor of φ_e, we let ϑ_e^* be the $\mathbb{D}$-inner function

$$\vartheta_e^*(z) := \overline{\vartheta_e \left(\frac{1}{\bar{z}} \right)}, \quad z \in \mathbb{D}. \tag{4.2.2}$$

From here, we form the space $\varphi \widetilde{K_{z\vartheta_e^*}}$ of analytic functions $\widehat{\mathbb{C}} \setminus \gamma$.

Remark 4.2.3. Technically speaking, $\widetilde{K_{z\vartheta_e^*}}$ is a space of analytic functions on $\widehat{\mathbb{C}} \setminus \mathbb{T}$ and not $\widehat{\mathbb{C}} \setminus \gamma$. However, since φ has an analytic continuation across $\mathbb{T} \setminus \gamma$, then so does ϑ_e (the $\mathbb{D}_e$-inner part of φ) [34, p. 78]. A version of Morera's theorem [16, p. 84] says that for each $f \in \widetilde{K_{z\vartheta_e^*}}$, the functions f_i and $\widetilde{f_i}$ are analytic continuations of each other across $\mathbb{T} \setminus \gamma$. Thus $\widetilde{K_{z\vartheta_e^*}}$, and hence $\varphi \widetilde{K_{z\vartheta_e^*}}$ can be considered to be a space of analytic functions on $\widehat{\mathbb{C}} \setminus \gamma$.

We will show that, in a sense, these spaces form the building blocks for every nearly invariant subspace.

To explain exactly what we mean here, let $\mathcal{N}$ be a nearly invariant subspace with greatest common $\widehat{\mathbb{C}} \setminus \gamma$-inner divisor equal to one and let E, ρ, F be the parameters in Theorem 3.1.2 and let φ be the extremal function for $\mathcal{N}$. Note, since $\rho := \varphi_i / \varphi_e$ (3.5.2), that $\mathcal{N}$ contains the nearly invariant subspace

$$\mathcal{N}_0 := \left\{ f \in H^2(\widehat{\mathbb{C}} \setminus \gamma) : \frac{f_i}{F} \in H^2(\mathbb{D}), \frac{f_i}{f_e} = \frac{\varphi_i}{\varphi_e} \text{ a.e. on } \gamma \right\}. \tag{4.2.4}$$

Observe how $\mathcal{N}_0$ is the intersection of the two closed sets

$$\mathcal{N} \quad \text{and} \quad \left\{ f \in H^2(\widehat{\mathbb{C}} \setminus \gamma) : \frac{f_i}{f_e} = \frac{\varphi_i}{\varphi_e} \text{ a.e. on } \gamma \right\}$$

and so $\mathcal{N}_0$ is closed. Moreover, $\varphi \in \mathcal{N}_0$. Our structure theorem here is the following.

Theorem 4.2.5. *Let $\mathcal{N}$ be a nearly invariant subspace of $H^2(\widehat{\mathbb{C}} \setminus \gamma)$ with greatest common $\widehat{\mathbb{C}} \setminus \gamma$-inner divisor equal to 1. Let φ be the extremal function for $\mathcal{N}$ and let ϑ_e be the $\mathbb{D}_e$-inner factor of φ_e. Then we have the following:*

1. The space $\mathcal{N}_0$ defined in (4.2.4) satisfies

$$\mathcal{N}_0 = \varphi \widetilde{K_{z\vartheta_e^*}}.$$

2. For any sequence $(A_n)_{n \geqslant 1}$ of closed subsets of $\gamma \setminus E$ with positive measure such that $m(A_n) \to m(\gamma \setminus E)$ we have

$$\mathcal{N} = \bigvee_{n=1}^{\infty} H^{\infty}(\widehat{\mathbb{C}} \setminus A_n) \mathcal{N}_0.$$

One of the keys to proving Theorem 4.2.5 will be this following special case.

Proposition 4.2.6. *Let $\mathcal{N}$ be a nearly invariant subspace of $H^2(\widehat{\mathbb{C}} \setminus \gamma)$ with greatest common $\widehat{\mathbb{C}} \setminus \gamma$-inner factor equal to 1 and with $E(\mathcal{N}) = \gamma$. If φ is the extremal function for $\mathcal{N}$ and ϑ_e is the $\mathbb{D}_e$-inner factor of $\varphi|\mathbb{D}_e$, then*

$$\mathcal{N} = \varphi \widetilde{K_{z\vartheta_e^*}}.$$

Proof. Recall from Corollary 3.3.1 the isometry $J : \mathcal{N} \to H^2(\mathbb{D}) \oplus L^2(\gamma, \omega)$ defined by

$$Jf = \left(\frac{f_i}{\varphi_i}, f_e - \frac{f_i}{\varphi_i} \varphi_e \right)$$

and note that since we are assuming that $E(\mathcal{N}) = \gamma$, then

$$J\mathcal{N} = \frac{1}{\varphi_i} \mathcal{N}|\mathbb{D} \oplus \{0\}.$$

We also know from Corollary 3.3.1 that the first component of $J\mathcal{N}$ is an S^*-invariant subspace of $H^2(\mathbb{D})$ which, since $\varphi \in \mathcal{N}$, contains the constants. From our discussion before, we know that

$$\frac{1}{\varphi_i} \mathcal{N}|\mathbb{D} = K_{z\vartheta}$$

for some $\mathbb{D}$-inner function ϑ. This says that

$$\mathcal{N} = \varphi \widetilde{K_{z\vartheta}}.$$

To finish the proof, we will show that $\vartheta = c\vartheta_e^*$ for some unimodular constant c.

From the Wold decomposition of $(J\mathcal{N})^{\perp}$, in particular Proposition 3.4.2 and (3.4.9), we see that

$$(J\mathcal{N})^{\perp} \cap (H^2(\mathbb{D}) \oplus \{0\}) = \left(\bigvee_{n=0}^{\infty} z^n a \right) \oplus \{0\}$$

for some $a \in H^{\infty}(\mathbb{D})$ for which, by Lemma 3.4.10, the function

$$z \mapsto \frac{a(z)}{\vartheta_e^*}$$

belongs to $N_0^+(\mathbb{D})$. This means that $a \in \vartheta_e^* z H^2(\mathbb{D})$ and so

$$(J\mathcal{N})^{\perp} \cap (H^2(\mathbb{D}) \oplus \{0\}) \subset \vartheta_e^* z H^2(\mathbb{D}) \oplus \{0\}.$$

But since the first component of $J\mathcal{N}$ is equal to $K_{z\vartheta} = H^2(\mathbb{D}) \ominus z\vartheta H^2(\mathbb{D})$, we observe that $z\vartheta H^2(\mathbb{D}) \subset \vartheta_e^* z H^2(\mathbb{D})$ and thus ϑ_e^* divides ϑ. On the other hand, for any $f \in \mathcal{N}$, we know that $f_i/\varphi_i \in K_{z\vartheta}$ and so, by our previous discussion in (4.1.3), f_i/φ_i has a pseudo-continuation to the function g/ϑ^* for some $g \in H^2(\mathbb{D}_e)$ (depending on f). We are also assuming that $E(\mathcal{N}) = \gamma$ and so

$$f_e - \frac{f_i}{\varphi_i} \varphi_e = 0 \quad \text{a.e. on } \gamma$$

and hence (since the above identity also holds on $\mathbb{T} \setminus \gamma$) f_i/φ_i has a pseudocontinuation to f_e/φ_e. Since pseudocontinuations are unique (Privalov's uniqueness theorem – [16, p. 13]) $g/\vartheta^* = f_e/\varphi_e$ and so

$$f_e = \frac{g}{\vartheta^*} \varphi_e.$$

It now must be the case that ϑ^* divides ϑ_e otherwise the restrictions to $\mathbb{D}_e$ of the functions in $\widetilde{K_{z\vartheta}}$ would have a common $\mathbb{D}_e$-inner factor – which is impossible since the greatest common $\widehat{\mathbb{C}} \setminus \gamma$-inner divisor of $\mathcal{N}$ is 1 (see Lemma 3.4.12). Thus ϑ divides ϑ_e^* and thus $\vartheta = c\vartheta_e^*$ for some unimodular constant c. $\qquad\square$

Thus we see that Theorem 4.2.5 works in some special cases. In order to prove the result for a general nearly invariant subspace $\mathcal{N}$, we need to take care of some technical details.

Lemma 4.2.7. *The Cauchy transform*

$$(C\mu)(z) := \int \frac{1}{\zeta - z} d\mu(\zeta), \quad z \in \widehat{\mathbb{C}} \setminus \gamma,$$

of a finite complex Borel measure on γ belongs to $N^+(\widehat{\mathbb{C}} \setminus \gamma)$.

Proof. Consider the arcs γ_n, $n \in \mathbb{N}$, with the same midpoint as γ but whose lengths satisfy $\ell(\gamma_n) = \ell(\gamma) + 1/n$. For each $n \in \mathbb{N}$ define

$$G_n := \widehat{\mathbb{C}} \setminus \bigcup \left\{ r\gamma_n : r > 0, |r - 1| \leqslant \frac{1}{n} \right\}$$

(see Figure 4.1).

On the arc $(1 - \frac{1}{n})\gamma_n$ note that

$$\omega_{G_n,0} \leqslant \omega_{\widehat{\mathbb{C}} \setminus (1-\frac{1}{n})\gamma_n, 0} \tag{4.2.8}$$

(see [23, p. 307] or [58, p. 102]). In a similar way,

$$\omega_{G_n,0} \leqslant \omega_{\widehat{\mathbb{C}} \setminus (1+\frac{1}{n})\gamma_n, 0} \tag{4.2.9}$$

on the arc $(1 + \frac{1}{n})\gamma_n$.

Manipulations with conformal mappings (see (2.3.13)) will show that

$$\omega_{\widehat{\mathbb{C}} \setminus (1-\frac{1}{n})\gamma_n, 0} \asymp \frac{1}{|z - a_n^-|^{1/2}|z - b_n^-|^{1/2}} ds, \tag{4.2.10}$$

where a_n^- and b_n^- are the endpoints of $(1 - \frac{1}{n})\gamma_n$. In a similar way,

$$\omega_{\widehat{\mathbb{C}} \setminus (1+\frac{1}{n})\gamma_n, 0} \asymp \frac{1}{|z - a_n^+|^{1/2}|z - b_n^+|^{1/2}} ds, \tag{4.2.11}$$

where a_n^+ and b_n^+ are the endpoints of $(1 + \frac{1}{n})\gamma_n$ (see Figure 4.1).

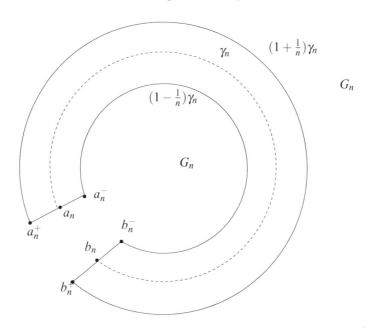

Figure 4.1: The region G_n.

If a_n and b_n are the endpoints of γ_n, we note that on the rays

$$L_n := \left\{ ra_n : |r - 1| \leqslant \frac{1}{n} \right\}, \quad R_n := \left\{ rb_n : |r - 1| \leqslant \frac{1}{n} \right\}$$

(the line segment connecting a_n^+ and a_n^- – respectively the line segment connecting b_n^+ and b_n^-) we have the estimate

$$\operatorname{dist}(z, \gamma) \geqslant \operatorname{dist}(z, \gamma_n) \geqslant C \min\{|z - a_n|, |z - b_n|\}, \tag{4.2.12}$$

with $C > 0$ independent of n.

With these estimates in place, consider the functions f_n on G_n defined by

$$f_n(z) := (z - a_n)(z - a_n^-)(z - a_n^+)(z - b_n)(z - b_n^-)(z - b_n^+)g(z),$$

where

$$g(z) := \frac{1}{z^6} \left((C\mu)(z) - \sum_{k=0}^{5} \frac{(C\mu)^{(k)}(0)z^k}{k!} \right).$$

For each $n \in \mathbb{N}$, f_n is bounded on G_n and thus $f_n \in H^{1/2}(G_n)$. Furthermore,

$$\int_{\partial G_n} |f_n|^{1/2} d\omega_{G_n,0} = \left(\int_{(1-\frac{1}{n})\gamma_n} + \int_{(1+\frac{1}{n})\gamma_n} + \int_{L_n} + \int_{R_n} \right) |f_n|^{1/2} d\omega_{G_n,0} =: \mathrm{I} + \mathrm{II} + \mathrm{III} + \mathrm{IV}.$$

Observe from (4.2.8) and (4.2.10) that

$$I \leqslant C \int_{(1-\frac{1}{n})\mathbb{T}} |g(z)|^{1/2} ds \leqslant C \|g\|_{H^{1/2}(\mathbb{D})}^{1/2}.$$

In a similar way, from (4.2.9) and (4.2.11), we have

$$II \leqslant C \|g\|_{H^{1/2}(\mathbb{D}_e)}^{1/2}.$$

From (4.2.12) the integrals III and IV are uniformly bounded in n. Putting this all together, we have

$$\sup_{n\in\mathbb{N}} \|f_n\|_{H^{1/2}(G_n)} < \infty. \qquad (4.2.13)$$

The regions G_n increase up to $\widehat{\mathbb{C}} \setminus \gamma$ as $n \to \infty$ and, for fixed k,

$$f_n(z) \to f(z) := (z+1)^3 (z+i)^3 g(z)$$

uniformly on G_k^- as $n \to \infty$. This means there is an $M > 0$, independent of n and k, so that

$$|f(z) - f_n(z)| \leqslant M, \quad z \in G_k^-, \quad n \geqslant k.$$

From here we get

$$|f(z)|^{1/2} \leqslant 2^{1/2}(M^{1/2} + |f_n(z)|^{1/2}), \quad z \in G_k^-, \quad n \geqslant k.$$

Since, by (4.2.13), the least harmonic majorant of $|f_n|^{1/2}$ on G_n at $z = 0$ is uniformly bounded in n, we see from the previous equation that $|f|^{1/2}$ has a harmonic majorant, and hence a least harmonic majorant, u_k on G_k and

$$\sup_k u_k(0) < \infty.$$

An application of Harnack's inequality says, for fixed $z \in G_{k_0}$, that

$$\sup_{k\geqslant k_0} u_k(z) < \infty.$$

Since u_k pointwise increases to a harmonic function u on $\widehat{\mathbb{C}} \setminus \gamma$, we see that $|f|^{1/2}$ has a harmonic majorant on $\widehat{\mathbb{C}} \setminus \gamma$, i.e., $f \in H^{1/2}(\widehat{\mathbb{C}} \setminus \gamma)$. Using the Nevanlinna theory it follows that $C\mu \in N^+(\widehat{\mathbb{C}} \setminus \gamma)$. $\qquad \square$

Remark 4.2.14. 1. Lemma 4.2.7 is considered a 'folklore' result. With a different proof and a little more effort, one can show that $C\mu \in H^p(\widehat{\mathbb{C}} \setminus \gamma)$ for all $0 < p < 1/2$. See [49, 74] for related results. In fact, a proof of Lemma 4.2.7 can be fashioned from [49]. We thank Dima Khavinson for pointing all this out to us.

2. If μ is a finite measure on $[0,1]$, one can adjust the proof of the previous corollary to show that $C\mu$ belongs to $N^+(\widehat{\mathbb{C}} \setminus [0,1])$. In fact if ν is a finite measure on ∂G, where $G = \mathbb{D} \setminus [0,1)$, we can write $\nu = \nu|\mathbb{T} + \nu|[0,1]$ and apply the fact that $C(\nu|\mathbb{T}) \in N^+(\mathbb{D})$ and the above observation to see that $C\nu \in N^+(G)$. We will make use of this several times later on.

Corollary 4.2.15. *Let f be analytic on $\widehat{\mathbb{C}} \setminus \gamma$ such that $f|\mathbb{D} \in H^1(\mathbb{D})$ and $f|\mathbb{D}_e \in H^1(\mathbb{D}_e)$. Then $f \in N^+(\widehat{\mathbb{C}} \setminus \gamma)$. Furthermore, if, for some $q \geqslant 1$,*

$$\int_\gamma (|f_i|^q + |f_e|^q)\,d\omega < \infty,$$

then $f \in H^q(\widehat{\mathbb{C}} \setminus \gamma)$.

Proof. The Cauchy integral formula says that $f - f(\infty)$ is the Cauchy transform of the finite measure

$$d\mu = (f_i - f_e)\frac{dz}{2\pi i}.$$

Now apply Lemma 4.2.7 and Proposition 2.4.10.   □

Here is the last technical lemma we need to prove Theorem 4.2.5.

Lemma 4.2.16. *Let Θ be a $\widehat{\mathbb{C}} \setminus \gamma$-inner function such that Θ_i is $\mathbb{D}$-outer and Θ_e is $\mathbb{D}_e$-outer. If a and b are the endpoints of γ, and Θ_a, Θ_b are the atomic singular $\widehat{\mathbb{C}} \setminus \gamma$-inner functions with singularity at a and b, then*

$$\Theta = B\Theta_a^t \Theta_b^s,$$

where $s, t \geqslant 0$ and B is a $\widehat{\mathbb{C}} \setminus \gamma$-Blaschke product – which can be equal to 1 – with zeros on $\mathbb{T} \setminus \gamma$.

Proof. The hypothesis that Θ_i and Θ_e are outer functions say that the zeros – if any – of the Blaschke factor B of Θ must lie on $\mathbb{T} \setminus \gamma$. Since Θ_i is bounded and $\mathbb{D}$-outer and has boundary values equal to 1 almost everywhere on γ, a version of the Schwarz reflection principle shows that Θ_i has an analytic continuation across γ. In a similar way, Θ_e has an analytic continuation across γ. This says that the $\widehat{\mathbb{C}} \setminus \gamma$-singular inner factor of Θ has a limit of modulus one when we approach any point in the interior of γ from both within $\mathbb{D}$ and from within $\mathbb{D}_e$. By a known result about limits of singular inner functions [34, p. 76] the singular $\widehat{\mathbb{C}} \setminus \gamma$-inner factor of Θ has no mass on the interior of γ.   □

Proof of Theorem 4.2.5. To show statement (1) we first notice that the extremal function for $\mathcal{N}_0$ is φ (since $\mathcal{N}_0 \subset \mathcal{N}$ and φ is the extremal function for $\mathcal{N}$) and $E(\mathcal{N}_0) = \gamma$. Once we show that the greatest common $\widehat{\mathbb{C}} \setminus \gamma$-inner factor of $\mathcal{N}_0$ is 1, we then use Proposition 4.2.6 to obtain the result. To this end, we observe that if ϑ_e is the inner factor of φ_e then

$$\vartheta_e^* \varphi_i \in H^2(\mathbb{D}) \quad \text{and} \quad \frac{\varphi_e}{\vartheta_e} \in H^2(\mathbb{D}_e).$$

The functions $\vartheta_e^* \varphi_i$ and φ_e/ϑ_e are analytic continuations of each other across $\mathbb{T} \setminus \gamma$ and so we can use Corollary 4.2.15 to see that the analytic function $\widetilde{\varphi}$ on $\widehat{\mathbb{C}} \setminus \gamma$ defined by

$$\widetilde{\varphi}(z) := \begin{cases} \vartheta_e^*(z)\varphi(z), & z \in \mathbb{D}; \\ \varphi(z)/\vartheta_e(z), & z \in \mathbb{D}_e. \end{cases} \tag{4.2.17}$$

belongs to $H^2(\widehat{\mathbb{C}} \setminus \gamma)$. By the definition of $\mathcal{N}_0$ in (4.2.4) it follows that $\widetilde{\varphi} \in \mathcal{N}_0$.

The greatest common $\widehat{\mathbb{C}} \setminus \gamma$-inner divisor Θ of $\mathcal{N}_0$ must divide both φ and $\widetilde{\varphi}$. Since $\varphi/\Theta \in N^+(\widehat{\mathbb{C}} \setminus \gamma)$, then $\varphi_i/\Theta_i \in N^+(\mathbb{D})$ [2] and since the greatest common $\widehat{\mathbb{C}} \setminus \gamma$-inner divisor of $\mathcal{N}$ is one, we know from (3.5.3) that φ_i is $\mathbb{D}$-outer and so Θ_i is $\mathbb{D}$-outer. On the other hand, since $\widetilde{\varphi}/\Theta \in N^+(\widehat{\mathbb{C}} \setminus \gamma)$, then $\widetilde{\varphi}_e/\Theta_e \in N^+(\mathbb{D}_e)$. But then, by the definition of $\widetilde{\varphi}_e$ in (4.2.17), Θ_e must divide the $\mathbb{D}_e$-outer part of φ_e and thus Θ_e must indeed be $\mathbb{D}_e$-outer. By Lemma 4.2.16 we have

$$\Theta = B\Theta_a^t \Theta_b^s.$$

To see that $B \equiv 1$, notice from Lemma 4.2.16 that the zeros of B – if any – must lie in $\mathbb{T} \setminus \gamma$. If there is indeed a zero z_0 of B in $\mathbb{T} \setminus \gamma$ then φ must also have this zero since Θ is the greatest common $\widehat{\mathbb{C}} \setminus \gamma$-inner divisor of $\mathcal{N}_0$ and $\varphi \in \mathcal{N}_0$. However, since the greatest common $\widehat{\mathbb{C}} \setminus \gamma$-inner divisor of $\mathcal{N}$ is 1, there is an $f \in \mathcal{N}$ with $f(z_0) \neq 0$. From Lemma 3.2.14, $f_i/\varphi_i \in H^2(\mathbb{D})$ which is impossible due to the zero of φ at z_0. In a similar way, one can see that if either s or t were positive, then $|\varphi|$ would go to zero too quickly along $\mathbb{T} \setminus \gamma$ for f_i/φ_i to belong to $H^2(\mathbb{D})$ for every $f \in \mathcal{N}$. This completes the proof of statement (1).

To see statement (2), first note that

$$\mathcal{N}_1 := \bigvee_n H^\infty(\widehat{\mathbb{C}} \setminus A_n)\mathcal{N}_0$$

is nearly invariant. Indeed, $\mathcal{N}_0$ is nearly invariant. Moreover, for $\lambda \notin Z(\mathcal{N}_1)$ (the common zero set of $\mathcal{N}_1$), $f,g \in \mathcal{N}_0$ with $g(\lambda) \neq 0$, and $u \in H^\infty(\widehat{\mathbb{C}} \setminus \gamma)$ we have

$$\frac{uf - \frac{uf}{g}(\lambda)g}{z - \lambda} = \frac{u - u(\lambda)}{z - \lambda}f + u(\lambda)\frac{f - \frac{f}{g}(\lambda)g}{z - \lambda}$$

which clearly belongs to $\mathcal{N}_1$. Thus for a dense set of functions $h \in \mathcal{N}_1$ we have

$$\frac{h - \frac{h}{g}(\lambda)g}{z - \lambda} \in \mathcal{N}_1$$

and the nearly invariance of $\mathcal{N}_1$ follows. Second, we apply our main theorem (Theorem 3.1.2) to see that

$$\mathcal{N}_1 = \left\{ f \in H^2(\widehat{\mathbb{C}} \setminus \gamma) : \frac{f_i}{F_1} \in H^2(\mathbb{D}), \frac{f_i}{f_e} = \rho_1 \text{ a.e. on } E_1 \right\}$$

---

[2] Indeed if $f \in N^+(\widehat{\mathbb{C}} \setminus \gamma)$, then $f = g/h$, where $g,h \in H^\infty(\widehat{\mathbb{C}} \setminus \gamma)$ and h is $\widehat{\mathbb{C}} \setminus \gamma$-outer. To show that $f_i \in N^+(\mathbb{D})$ it suffices to show that h_i is $\mathbb{D}$-outer. Let $\mathcal{N}$ be a closure of $hH^\infty(\widehat{\mathbb{C}} \setminus \gamma)$ and note, since h is $\widehat{\mathbb{C}} \setminus \gamma$-outer, that $\mathcal{N} = H^2(\widehat{\mathbb{C}} \setminus \gamma)$. Now follow the end of the proof of Lemma 3.4.12.

for some $\mathbb{D}$-outer function F_1, some measurable ρ_1 on γ, and some measurable set $E_1 \subset \gamma$. The space $\mathcal{N}$ is described by the parameters F, ρ, E in a similar way. Since $\varphi \in \mathcal{N}_1$ and since, from (3.5.2), ρ_1 and ρ are formed in the same way from φ, we see that $\rho = \rho_1$. Using (3.5.1) we see, in the same way, that F_1 can be taken to be equal to F. Since $\mathcal{N}_1 \subset \mathcal{N}$ (Corollary 3.6.2) we see that $E_1 \supset E$. Indeed, $\mathcal{N}_1$ contains a dense set of functions f for which $f_i/f_e = \rho$ almost everywhere on E. To see that $E_1 = E$ almost everywhere, we proceed as follows: Suppose that $m(E_1 \setminus E) > 0$, then, by the definition of the A_n's, $m(A_n \cap (E_1 \setminus E)) > 0$ for some n. Let $u \in H^\infty(\widehat{\mathbb{C}} \setminus A_n)$ be the function from (3.6.1), i.e., $u_i/u_e \neq 1$ on some compact subset of A_n of positive measure. Then $f := u\varphi \in \mathcal{N}_1$ but

$$\frac{f_i}{f_e} = \frac{u_e}{u_i} \frac{\varphi_i}{\varphi_e} = \frac{u_e}{u_i} \rho$$

and this last function is not equal to ρ almost everywhere on E_1, a contradiction. $\qquad\square$

In certain special cases, we have a refinement of Theorem 4.2.5.

Corollary 4.2.18. *Let $\mathcal{N}$ be a nearly invariant subspace of $H^2(\widehat{\mathbb{C}} \setminus \gamma)$ with greatest common $\widehat{\mathbb{C}} \setminus \gamma$-inner divisor equal to 1. If φ_e is $\mathbb{D}_e$-outer, then φ is $\widehat{\mathbb{C}} \setminus \gamma$-outer and furthermore, for any sequence $(A_n)_{n \geqslant 1}$ of closed subsets of $\gamma \setminus E$ with positive measure such that $m(A_n) \to m(\gamma \setminus E)$ we have*

$$\mathcal{N} = \bigvee_{n=1}^{\infty} H^\infty(\widehat{\mathbb{C}} \setminus A_n)\varphi. \qquad (4.2.19)$$

Proof. In this case $\vartheta_e \equiv 1$ and so $\widetilde{K_{z\vartheta_e^*}} = \mathbb{C}$. The identity in (4.2.19) now follows from Theorem 4.2.5. Since the greatest common $\widehat{\mathbb{C}} \setminus \gamma$-inner factor of $\mathcal{N}$ is 1, (4.2.19) shows that φ is $\widehat{\mathbb{C}} \setminus \gamma$-outer. $\qquad\square$

Corollary 4.2.20. *Let $\mathcal{N}$ be a nearly invariant subspace of $H^2(\widehat{\mathbb{C}} \setminus \gamma)$ which contains the constants and let F, ρ, and E be the parameters in Theorem 3.1.2.*

1. *The outer function F can be chosen to be equal to 1 and the function ρ is equal to one almost everywhere, i.e.,*

$$\mathcal{N} = \left\{ f \in H^2(\widehat{\mathbb{C}} \setminus \gamma) : \frac{f_i}{f_e} = 1 \ a.e. \ on \ E \right\}. \qquad (4.2.21)$$

2. *For any sequence $(A_n)_{n \geqslant 1}$ of closed subsets of $\gamma \setminus E$ with positive measure such that $m(A_n) \to m(\gamma \setminus E)$, we have*

$$\mathcal{N} = \bigvee_{n=1}^{\infty} H^\infty(\widehat{\mathbb{C}} \setminus A_n). \qquad (4.2.22)$$

3. *If E is open in γ and $E^c = \gamma \setminus E$, then*

$$\mathcal{N} = clos_{H^2(\widehat{\mathbb{C}} \setminus \gamma)} H^\infty(\widehat{\mathbb{C}} \setminus E^c) \qquad (4.2.23)$$

and this set is equal to

$$\left\{ f \in H^2(\widehat{\mathbb{C}} \setminus \gamma) : f \text{ extends analytically across } E \right\}. \qquad (4.2.24)$$

Proof. Consider the description of $\mathcal{N}$ via Theorem 3.1.2 with the parameters ρ, E, F. Because $\mathcal{N}$ contains the constants and thus $\varphi \equiv 1$ is the extremal function for $\mathcal{N}$ (see Remark 4.2.1), we see from (3.5.2) that $\rho = \varphi_i/\varphi_e = 1$. From (3.5.1), F can be taken to be the $\mathbb{D}$-outer function such that

$$|F|^2 = \frac{1}{1 + w\chi_\gamma}, \quad \text{a.e.}$$

From here one can see that

$$\left\{ f \in H^2(\widehat{\mathbb{C}} \setminus \gamma) : \frac{f_i}{F} \in H^2(\mathbb{D}) \right\} = H^2(\widehat{\mathbb{C}} \setminus \gamma)$$

and so we can take $F \equiv 1$. This proves statement (1).

Statement (2) follows directly from Corollary 4.2.18 since $\varphi \equiv 1$.

To prove statement (3) we see that (4.2.23) follows from statement (4.2.22) with $A_n = E^c$ for all n. From (4.2.22) and Morera's theorem (see [34, p. 95] or Proposition 6.2.8 below),

$$\mathcal{N} \subset \left\{ f \in H^2(\widehat{\mathbb{C}} \setminus \gamma) : f \text{ extends analytically across } E \right\}.$$

The reverse inclusion follows from (4.2.21). $\qquad \square$

Chapter 5

Nearly invariant and de Branges spaces

5.1 de Branges spaces

It turns out that we can also describe the nearly invariant subspaces of $H^2(\widehat{\mathbb{C}} \setminus \gamma)$ in terms of a de Branges-type space on $\widehat{\mathbb{C}} \setminus \mathbb{T}$. First let us review the well-known de Branges spaces on $\mathbb{C} \setminus \mathbb{R}$. We follow [25, p. 9–12]. Let Ψ be an analytic function on the upper half plane $\mathbb{C}_+ = \{\Im z > 0\}$ such that $\Re\Psi \geqslant 0$. The classical Herglotz theorem [25, p. 7] says that there is a non-negative measure μ on $\mathbb{R}$ and a non-negative number p such that

$$\Re\Psi(x+iy) = py + \frac{1}{\pi} \int_{-\infty}^{\infty} \frac{y}{(t-x)^2 + y^2} d\mu(t), \quad x+iy \in \mathbb{C}_+. \qquad (5.1.1)$$

The reader will recognize the above integral as the Poisson integral of μ. Extend Ψ to the lower half plane so that

$$\Psi(z) = -\overline{\Psi(\bar{z})}, \quad z = x+iy, \quad y < 0.$$

A theorem of de Branges [25, p. 9] says that there exists a unique Hilbert space $\mathcal{L}(\Psi)$ of analytic functions on $\mathbb{C} \setminus \mathbb{R}$ such that for each fixed $w \in \mathbb{C} \setminus \mathbb{R}$, the function

$$z \mapsto \frac{\Psi(z) + \overline{\Psi(w)}}{\pi i (\bar{w} - z)} \qquad (5.1.2)$$

belongs to $\mathcal{L}(\Psi)$ and

$$F(w) = \left\langle F(z), \frac{\Psi(z) + \overline{\Psi(w)}}{\pi i (\bar{w} - z)} \right\rangle_{\mathcal{L}(\Psi)} \quad \forall F \in \mathcal{L}(\Psi). \qquad (5.1.3)$$

The previous identity says that the functions in (5.1.2) are the reproducing kernel functions for $\mathcal{L}(\Psi)$. Furthermore, if μ is the measure from (5.1.1), the linear transformation

$$f \mapsto \frac{1}{\pi i} \int_{-\infty}^{\infty} \frac{f(t)}{t - z} d\mu(t) \qquad (5.1.4)$$

maps $L^2(\mu)$ isometrically into $\mathcal{L}(\Psi)$ and the orthogonal complement of the range of this transformation contains only constant functions. For example, if $p = 0$ in (5.1.1), this map is onto.

Let us create a de Branges-type space of analytic functions on $\widehat{\mathbb{C}} \setminus \mathbb{T}$. For this, let Φ be analytic on $\mathbb{D}$ with non-negative real part and extend Φ to $\widehat{\mathbb{C}} \setminus \mathbb{T}$ by

$$\Phi(z) = -\overline{\Phi(1/\overline{z})}, \quad z \in \mathbb{D}_e.$$

By the change of variable

$$z \mapsto i \frac{1+z}{1-z} \qquad (5.1.5)$$

(which maps $\mathbb{D}$ onto $\mathbb{C}_+$) in (5.1.2) and (5.1.3), we create a unique reproducing kernel Hilbert space $\widetilde{\mathcal{L}}(\Phi)$ of analytic functions on $\widehat{\mathbb{C}} \setminus \mathbb{T}$ with kernel function

$$k^{\Phi}(w,z) = (1 - \overline{w})(1 - z) \frac{\Phi(z) + \overline{\Phi(w)}}{2\pi(1 - \overline{w}z)}, \quad z, w \in \widehat{\mathbb{C}} \setminus \mathbb{T}. \qquad (5.1.6)$$

That is to say,

$$\langle F(\cdot), k^{\Phi}(w, \cdot) \rangle_{\widetilde{\mathcal{L}}(\Phi)} = F(w), \quad F \in \widetilde{\mathcal{L}}(\Phi), \quad w \in \widehat{\mathbb{C}} \setminus \mathbb{T}.$$

Applying the change of variable in (5.1.5) along with the integral change of variable

$$t \mapsto -i \frac{1+\overline{\zeta}}{1-\overline{\zeta}}$$

to the integral in (5.1.4), we create an operator

$$V : L^2(\widetilde{\mu}) \to \widetilde{\mathcal{L}}(\Phi), \quad (Vf)(z) := \frac{1}{2\pi}(1 - z) \int_{\mathbb{T}} \frac{(1 - \overline{\zeta})f(\zeta)}{1 - \overline{\zeta}z} d\widetilde{\mu}(\zeta).$$

Here $\widetilde{\mu}$ is the pullback measure on $\mathbb{T}$ formed from μ (on $\mathbb{R}$) via the above change of variable. This operator V is an isometry and the orthogonal complement of the range of V contains only constant functions.

5.2   de Branges spaces and nearly invariant subspaces

We will now associate each nearly invariant subspace of $H^2(\widehat{\mathbb{C}} \setminus \gamma)$ with one of these de Branges-type space $\widetilde{\mathcal{L}}(\Phi)$. All of our results on nearly invariant subspaces so far do not depend on the fact that $\gamma = \{e^{it} : -\pi/2 \leqslant t \leqslant \pi\}$ and still hold when γ is any proper sub-arc of $\mathbb{T}$. We will also assume, without loss of generality, that

$$1 \notin \gamma.$$

For our nearly invariant subspace $\mathcal{N}$ of $H^2(\widehat{\mathbb{C}} \setminus \gamma)$, we let, as in (3.2.2) and (3.2.3), φ and ψ denote the normalized reproducing kernel functions for $\mathcal{N}$ at $z = 0$ and $z = \infty$. From Corollary 3.2.9 (statement (2)) we know that $|\varphi(1)| = |\psi(1)|$. Multiplying by an appropriate unimodular constant, which will not change the fact that φ and ψ satisfy (3.2.1) and (3.2.3), we can assume that

$$\varphi(1) = \psi(1).$$

A computation using statements (1) and (2) of Corollary 3.2.9 along with the fact that

$$z \mapsto \frac{1+z}{1-z}$$

maps $\mathbb{D}$ onto $\{\Re z > 0\}$ will show that

$$\Phi := \frac{1 + \frac{z\psi}{\varphi}}{1 - \frac{z\psi}{\varphi}} \tag{5.2.1}$$

is an analytic function on $\widehat{\mathbb{C}} \setminus \mathbb{T}$ satisfying

$$\Re\Phi > 0 \quad \text{and} \quad \Phi(z) = -\overline{\Phi(1/\overline{z})}.$$

Thus from above, we can form the de Branges-type space $\widetilde{\mathcal{L}}(\Phi)$ along with the associated reproducing kernel k^Φ in (5.1.6). Recall from (3.2.7) that the reproducing kernel for $\mathcal{N}$ is

$$k_\lambda^{\mathcal{N}}(z) = \frac{\overline{\varphi(\lambda)}\varphi(z) - \overline{\lambda}z\overline{\psi(\lambda)}\psi(z)}{1 - \overline{\lambda}z}, \quad z \neq 1/\overline{\lambda}.$$

The next lemma relates the kernels $k_\lambda^{\mathcal{N}}$ and k^Φ.

Lemma 5.2.2. *If $\mathcal{N}$ is a nearly invariant subspace of $H^2(\widehat{\mathbb{C}} \setminus \gamma)$ with reproducing kernel $k_\lambda^{\mathcal{N}}$ and Φ is given by (5.2.1), then*

$$k_\lambda^{\mathcal{N}}(z) = \frac{1}{2}|\varphi(1)|^2 \overline{k_1^{\mathcal{N}}(\lambda)} k_1^{\mathcal{N}}(z) k^\Phi(\lambda, z).$$

Proof. From the definition of Φ from (5.2.1) we get

$$\frac{\Phi - 1}{\Phi + 1} = \frac{z\psi}{\varphi}.$$

This means that

$$k_\lambda^{\mathcal{N}}(z) = \overline{\varphi(\lambda)}\varphi(z)\frac{1 - \frac{\overline{\Phi(\lambda)}-1}{\overline{\Phi(\lambda)}+1}\frac{\Phi(z)-1}{\Phi(z)+1}}{1 - \overline{\lambda}z}$$

$$= \frac{2\overline{\varphi(\lambda)}\varphi(z)}{(\overline{\Phi(\lambda)}+1)(\Phi(z)+1)}\frac{\Phi(z)+\overline{\Phi(\lambda)}}{1-\overline{\lambda}z}.$$

Using (5.2.1) we observe that

$$\frac{\varphi}{\Phi+1} = \frac{\varphi - z\psi}{2}.$$

Using the definition of $k_1^{\mathcal{N}}$ and our assumption that $\varphi(1) = \psi(1)$ we see that

$$\varphi - z\psi = (1-z)\varphi(1)k_1^{\mathcal{N}}.$$

Now combine these last two identities with the above computation for $k_\lambda^{\mathcal{N}}$ to get

$$k_\lambda^{\mathcal{N}}(z) = \frac{1}{2}|\varphi(1)|^2\overline{k_1^{\mathcal{N}}(\lambda)}k_1^{\mathcal{N}}(z)(1-z)(1-\overline{\lambda})\frac{\Phi(z)+\overline{\Phi(\lambda)}}{1-\overline{\lambda}z}$$

$$= \frac{1}{2}|\varphi(1)|^2\overline{k_1^{\mathcal{N}}(\lambda)}k_1^{\mathcal{N}}(z)k^{\Phi}(\lambda,z).$$ $\qquad\square$

Our main result relating $\mathcal{N}$ with $\widetilde{\mathcal{L}}(\Phi)$ is the following.

Theorem 5.2.3. *With the assumptions above we have*

$$\mathcal{N} = k_1^{\mathcal{N}}\widetilde{\mathcal{L}}(\Phi).$$

Moreover, the operator

$$f \mapsto \frac{\varphi(1)}{\sqrt{2}}k_1^{\mathcal{N}}f$$

is an isometry from $\widetilde{\mathcal{L}}(\Phi)$ onto $\mathcal{N}$.

Proof. Given

$$f = \sum_j c_j k_{\lambda_j}^{\mathcal{N}},$$

a finite linear combination of reproducing kernel functions for $\mathcal{N}$ (which are dense in $\mathcal{N}$), define

$$Tf := \sum_j \frac{c_j}{\sqrt{2}}\overline{\varphi(1)k_1^{\mathcal{N}}(\lambda_j)}k_{\lambda_j}^{\Phi}$$

and observe that

$$\langle Tf, Tf \rangle_{\widetilde{\mathcal{L}}(\Phi)} = \sum_{j,l} \frac{c_j\overline{c_l}}{2}|\varphi(1)|^2\overline{k_1^{\mathcal{N}}(\lambda_j)}k_1^{\mathcal{N}}(\lambda_l)k^{\Phi}(\lambda_j,\lambda_l)$$

$$= \sum_{j,l} c_j\overline{c_l}k_{\lambda_j}^{\mathcal{N}}(\lambda_l) \quad \text{(by Lemma 5.2.2)}$$

$$= \langle f, f \rangle_{H^2(\widehat{\mathbb{C}}\backslash\gamma)}.$$

By standard arguments, we can extend T to a unitary operator from $\mathcal{N}$ onto $\widetilde{\mathcal{L}}(\Phi)$. Finally, for $F \in \widetilde{\mathcal{L}}(\Phi)$ and $\lambda \in \widehat{\mathbb{C}} \setminus \mathbb{T}$, we have

$$\langle T^*F, k_\lambda^{\mathcal{N}} \rangle_{H^2(\widehat{\mathbb{C}} \setminus \gamma)} = \langle F, T k_\lambda^{\mathcal{N}} \rangle_{\widetilde{\mathcal{L}}(\Phi)}$$

$$= \left\langle F, \frac{1}{\sqrt{2}} \overline{\varphi(1) k_1^{\mathcal{N}}(\lambda)} k^\Phi(\lambda, \cdot) \right\rangle_{\widetilde{\mathcal{L}}(\Phi)}$$

$$= \frac{\varphi(1)}{\sqrt{2}} k_1^{\mathcal{N}}(\lambda) F(\lambda).$$

This says that

$$T^*F = \frac{\varphi(1)}{\sqrt{2}} k_1 F$$

and is an isometry. $\qquad\square$

Chapter 6

Invariant subspaces of the slit disk

6.1 First description of the invariant subspaces

In this section we use our main theorem about nearly invariant subspaces (Theorem 3.1.2) and the conformal map $\alpha : G \to \widehat{\mathbb{C}} \setminus \gamma$ from (2.3.8) to give a full description of the invariant subspaces (under $Sf = zf$) of $H^2(G)$. Let us get started with a few preliminary observations.

Proposition 6.1.1. *A subspace* $\mathcal{M} \subset H^2(G)$ *is invariant if and only if* $\mathcal{M}$ *is* $H^\infty(\mathbb{D})$-*invariant, i.e.,* $g\mathcal{M} \subset \mathcal{M}$ *for every* $g \in H^\infty(\mathbb{D})$.

Proof. One direction of the argument is obvious. For the other, suppose that $f \in \mathcal{M}$ and $\phi \in H^\infty(\mathbb{D})$. Let $(\phi_n)_{n \geqslant 1}$ be a sequence of analytic polynomials such that $\phi_n \to \phi$ weak-$*$ in $H^\infty(\mathbb{D})$,[1] i.e., $\phi_n \to \phi$ pointwise in $\mathbb{D}$ and the sup-norms of ϕ_n are uniformly bounded in n. Since $\phi_n f \to \phi f$ pointwise in G and the $H^2(G)$-norms of $\phi_n f$ are uniformly bounded, it follows that $\phi_n f \to \phi f$ weakly in $H^2(G)$ [14, p. 272].

Note that $\phi_n f \in \mathcal{M}$ (since $\mathcal{M}$ is invariant) and ϕf belongs to the weak-closure of $\mathcal{M}$. By standard functional analysis, the weak-closure of $\mathcal{M}$ is equal to its norm closure [19, p. 129] and so $\phi f \in \mathcal{M}$. $\qquad\square$

Proposition 6.1.2. *A non-zero subspace* $\mathcal{M} \subset H^2(G)$ *is* $H^\infty(G)$-*invariant if and only if* $\mathcal{M} = \Theta H^2(G)$ *for some* G-*inner function* Θ.

Proof. Recall from (2.1.3) that if ϕ_G is a conformal map from $\mathbb{D}$ onto G, the composition operator

$$C_{\phi_G} f = f \circ \phi_G$$

---

[1] The sequence of Cesàro polynomials of ϕ will work [45, p. 19].

is a unitary operator from $H^2(G)$ onto $H^2(\mathbb{D})$. Now on to the proof. One direction is clear. For the other, suppose $\mathcal{M} \neq \{0\}$ and $H^\infty(G)$-invariant. Then $C_{\phi_G}\mathcal{M}$ is a non-zero $H^\infty(\mathbb{D})$-invariant subspace of $H^2(\mathbb{D})$. By Beurling's classical theorem which characterizes the invariant subspaces of $H^2(\mathbb{D})$ [34, p. 82], $C_{\phi_G}\mathcal{M} = IH^2(\mathbb{D})$, where I is a $\mathbb{D}$-inner function. Finally, notice that $\mathcal{M} = (I \circ \phi_G^{-1})H^2(G)$ and, by definition, $I \circ \phi_G^{-1}$ is G-inner.   □

Corollary 6.1.3. *An invariant subspace $\mathcal{M} \subset H^2(G)$ is nearly invariant.*

Proof. By Proposition 3.1.1, it suffices to show that whenever $f, g \in \mathcal{M}$ and $h \in \mathcal{M}^\perp$,

$$\left\langle \frac{f - \dfrac{f}{g}(a)g}{z - a}, h \right\rangle = 0 \quad \forall a \in G \setminus (Z(\mathcal{M}) \cup g^{-1}(\{0\})). \tag{6.1.4}$$

Recall that $Z(\mathcal{M})$ is the set of common zeros of $\mathcal{M}$. Let $W(a)$ be equal to the meromorphic function on G defined by the left-hand side of the above equation. Notice that W can be written in the form

$$W(a) = \int_{\mathbb{T}} \frac{d\mu_1(\zeta)}{\zeta - a} + \int_{[0,1]} \frac{d\mu_2(x)}{x - a} + \frac{f}{g}(a)\int_{\mathbb{T}} \frac{d\mu_3(\zeta)}{\zeta - a} + \frac{f}{g}(a)\int_{[0,1]} \frac{d\mu_4(x)}{x - a}.$$

One can argue, using some ideas from Remark 4.2.14, that W is in the Nevanlinna class of G and hence has finite non-tangential limits almost everywhere on ∂G.

   Since $\mathcal{M}$ is invariant, we have

$$\left\langle \frac{f - \dfrac{f}{g}(a)g}{z - b}, h \right\rangle = 0 \quad \forall |b| > 1.$$

Thus

$$W(a) = \left\langle \frac{f - \dfrac{f}{g}(a)g}{z - a}, h \right\rangle - \left\langle \frac{f - \dfrac{f}{g}(a)g}{z - b}, h \right\rangle, \quad a \in G \setminus g^{-1}(\{0\}), |b| > 1.$$

For $r \in (0, 1)$ and $\zeta \in \mathbb{T}$, let $a = r\zeta$ and $b = \zeta/r$. Apply Fatou's jump theorem (Theorem 3.3.3) to see that

$$\lim_{r \to 1^-} W(r\zeta) = 0 \quad \text{a.e. } \zeta \in \mathbb{T}. \tag{6.1.5}$$

Observe in the inner product how the contribution from the parts of the integral on the slit $[0, 1]$ cancel out in the limit.

   Using the well-known fact that if a Nevanlinna function has vanishing radial boundary values on a set of positive measure, then this function must vanish identically, along with (6.1.5), we see that $W \equiv 0$. This proves (6.1.4).   □

   We are now ready for our first description of the invariant subspaces of $H^2(G)$. We will see another description of them later on in Theorem 6.2.1.

Corollary 6.1.6. *Let* $\mathcal{M}$ *be a non-trivial invariant subspace of* $H^2(G)$ *with greatest common* G-*inner divisor* $\Theta_{\mathcal{M}}$. *Then there exists a* $\mathbb{D}_+$-*outer function* F, *a measurable set* $E \subset [0,1]$, *and a measurable function* $\rho : [0,1] \to \mathbb{C}$ *such that*

$$\mathcal{M} = \Theta_{\mathcal{M}} \cdot \left\{ f \in H^2(G) : \frac{f|\mathbb{D}_+}{F} \in H^2(\mathbb{D}_+), f^+ = \rho f^- \text{ a.e. on } E \right\}.$$

Proof. Consider the conformal map $\alpha : G \to \widehat{\mathbb{C}} \setminus \gamma$ from (2.3.8). Recall the unitary operator $C_{\alpha^{-1}} h = h \circ \alpha^{-1}$ from $H^2(G)$ onto $H^2(\widehat{\mathbb{C}} \setminus \gamma)$. We will now use Proposition 3.1.1 to show that $C_{\alpha^{-1}} \mathcal{M}$ is nearly invariant. Indeed if $\lambda \in \widehat{\mathbb{C}} \setminus \gamma$ (and not in the common zero set of $C_{\alpha^{-1}} \mathcal{M}$) and $f \in \mathcal{M}$ with $(f \circ \alpha^{-1})(\lambda) = 0$, we need to show that

$$\frac{f \circ \alpha^{-1}}{z - \lambda} \in C_{\alpha^{-1}} \mathcal{M}.$$

But this is equivalent to showing that

$$\frac{f}{\alpha - \lambda} \in \mathcal{M}.$$

However,

$$\frac{f}{\alpha - \lambda} = \frac{z - \alpha^{-1}(\lambda)}{\alpha - \lambda} \frac{f}{z - \alpha^{-1}(\lambda)}.$$

The second factor belongs to $\mathcal{M}$, since $\mathcal{M}$ is nearly invariant (Corollary 6.1.3). The first factor belongs to $H^\infty(\mathbb{D})$ [2] and $\mathcal{M}$ is $H^\infty(\mathbb{D})$-invariant (Proposition 6.1.1). Thus $f/(\alpha - \lambda) \in \mathcal{M}$ and so $C_{\alpha^{-1}} \mathcal{M}$ is nearly invariant.

Since $\mathcal{M}$ is $H^\infty(\mathbb{D})$-invariant,

$$\alpha([0,1]) = \gamma'' := \{e^{it} : 3\pi/2 \leqslant t \leqslant 2\pi\},$$

then $C_{\alpha^{-1}} \mathcal{M}$ is not only nearly invariant but is also $H^\infty(\widehat{\mathbb{C}} \setminus \gamma')$-invariant, where $\gamma' := \{e^{it} : 0 \leqslant t \leqslant \pi\}$. By Theorem 3.1.2,

$$C_{\alpha^{-1}} \mathcal{M} = \Theta \cdot \left\{ f \in H^2(\widehat{\mathbb{C}} \setminus \gamma) : \frac{f_i}{F} \in H^2(\mathbb{D}), f_i = \rho f_e \text{ a.e. on } E \right\}$$

for some $\widehat{\mathbb{C}} \setminus \gamma$-inner Θ, some $\mathbb{D}$-outer F, some measurable $E \subset \gamma''$ (note that $C_{\alpha^{-1}} \mathcal{M}$ is $H^\infty(\widehat{\mathbb{C}} \setminus \gamma)$-invariant), and some measurable $\rho : E \to \mathbb{C}$. Thus

$$\mathcal{M} = \widetilde{\Theta} \cdot \left\{ f \in H^2(G) : \frac{f|\mathbb{D}_+}{\widetilde{F}} \in H^2(\mathbb{D}_+), f^+ = \widetilde{\rho} f^- \text{ a.e. on } \widetilde{E} \right\},$$

where $\widetilde{\Theta} := \Theta \circ \alpha$ is G-inner, $\widetilde{F} := F \circ \alpha$ is $\mathbb{D}_+$-outer (since $\alpha(\mathbb{D}_+) = \mathbb{D}$), $\widetilde{E} := \alpha^{-1}(E) \subset [0,1]$ (since $E \subset \gamma''$ and $\alpha([0,1]) = \gamma''$), and $\widetilde{\rho} = \rho \circ \alpha$. $\qquad \square$

---

[2] See (2.3.8) and notice how α is analytic on $\mathbb{D} \setminus \{i(1 - \sqrt{2})\}$ with a simple pole at $i(1 - \sqrt{2})$.

Remark 6.1.7. By Corollary 3.6.3, the parameters $\widetilde{\Theta}$, $\widetilde{\rho}$, and $\widetilde{E}$ are (essentially) unique.

We also have the following version of Corollary 4.2.20.

Corollary 6.1.8. *Let A be a closed subset of $[0,1]$. For $f \in H^2(G)$ the following are equivalent.*

1. $f \in clos_{H^2(G)} H^\infty(\mathbb{D} \setminus A)$;

2. $f^+ = f^-$ almost everywhere on $[0,1] \setminus A$;

3. f has an analytic continuation across $[0,1) \setminus A$.

6.2   Second description of the invariant subspaces

The description of the invariant subspaces of $H^2(G)$ in Corollary 6.1.6 depends on the somewhat unnatural use of the $\mathbb{D}_+$-outer function F. This next result is an alternate, and perhaps more natural, description.

For $\varepsilon \in (0,1)$ let

$$G_\varepsilon := \mathbb{D} \setminus [-\varepsilon, 1).$$

Theorem 6.2.1. *For an invariant subspace $\mathcal{M}$ of $H^2(G)$, let $\Theta_{\mathcal{M}}$, E, and ρ be as in Corollary 6.1.6. Then for every $\varepsilon \in (0,1)$, there is a G_ε-outer function F_ε such that*

$$\mathcal{M} = \Theta_{\mathcal{M}} \cdot \left\{ f \in H^2(G) : \frac{f}{F_\varepsilon} \in H^2(G_\varepsilon),\ f^+ = \rho f^- \text{ a.e. on } E \right\}.$$

The proof of this theorem needs quite a few preliminaries. Let ω, ω_ε, and ω_+ be the harmonic measures for G (respectively G_ε and $\mathbb{D}_+$) at some common point in $\mathbb{D}_+$. If $\Omega = G$ (or G_ε or $\mathbb{D}_+$), we have

$$d\omega \asymp w_\Omega \frac{ds}{2\pi} \quad \text{and} \quad w_\Omega \asymp |\psi'_\Omega|, \tag{6.2.2}$$

where ψ is a conformal map from Ω onto $\mathbb{D}$ and ds is arc length measure on $\partial\Omega$. See this from (2.3.5) for G and G_ε and (2.2.3) for $\mathbb{D}_+$. We will use the notation

$$w := w_G, \quad w_\varepsilon := w_{G_\varepsilon}, \quad w_+ := w_{\mathbb{D}_+}.$$

Remark 6.2.3. If ω is harmonic measure for G, we will assume that $\psi^{-1}(0) \in \mathbb{D}_+$ and

$$\omega = \omega_{\psi^{-1}(0)}$$

and so by (2.3.5)

$$d\omega = |\psi'| \frac{ds}{2\pi}. \tag{6.2.4}$$

Having ω precisely as in (6.2.4) will become important in one of the technical lemmas below (see Lemma 6.2.14).

Recall, from our discussion of the estimates of harmonic measure in Chapter 2, that if η is one of the corners of $\partial\Omega$ with opening θ $(0 < \theta \leqslant 2\pi)$, then

$$|\psi'(\xi)| \asymp |\xi - \eta|^{\frac{\pi}{\theta}-1}, \quad \xi \approx \eta. \tag{6.2.5}$$

Our first technical lemma is standard [23, p. 307] [58, p. 102].

Lemma 6.2.6. *For each* $\varepsilon \in (0,1)$, *$\omega_\varepsilon \leqslant \omega$ on ∂G.*

Lemma 6.2.7. *1.* $\log w \in L^1(\partial G, \omega)$.

2. For each $\varepsilon \in (0,1)$,

$$\int_{\mathbb{T}_+ \cup [-\varepsilon,1]} \left| \log \frac{w_\varepsilon}{w_+} \right| w_\varepsilon ds < \infty,$$

where $\mathbb{T}_+ := \{e^{i\theta} : 0 \leqslant \theta \leqslant \pi\}$.

Proof. From (6.2.2) and (6.2.5) we have

$$
\begin{array}{llll}
w_+ \asymp |\xi + 1|, & w_\varepsilon \asymp 1, & w \asymp 1 & \text{for } \xi \approx -1; \\
w_+ \asymp |\xi - 1|, & w_\varepsilon \asymp |\xi - 1|, & w \asymp |\xi - 1| & \text{for } \xi \approx 1; \\
w_+ \asymp 1, & w_\varepsilon \asymp |\xi + \varepsilon|^{-1/2}, & w \asymp 1 & \text{for } \xi \approx -\varepsilon; \\
w_+ \asymp 1, & w_\varepsilon \asymp 1, & w \asymp |\xi|^{-1/2} & \text{for } \xi \approx 0.
\end{array}
$$

Furthermore, if one stays away from the points $\xi = 1, -1, 0, -\varepsilon$, the functions w_+, w_ε, w are continuous and bounded away from zero. The result follows. $\qquad\square$

We will also make use of the following Morera-type theorem [34, p. 95]. Recall the definition of the Hardy-Smirnov classes E^1 from (2.4.8).

Proposition 6.2.8. *Suppose* $f_1 \in E^1(\mathbb{D}_+)$ *and* $f_2 \in E^1(\mathbb{D}_-)$ *with*

$$\lim_{y \to 0^+} f_1(x + iy) = \lim_{y \to 0^-} f_2(x + iy)$$

almost everywhere on $[-1, 1]$. *Then the function*

$$g(z) := \begin{cases} f_1(z), & z \in \mathbb{D}_+; \\ f_2(z), & z \in \mathbb{D}_- \end{cases}$$

has an analytic continuation to $\mathbb{D}$.

Proof. Using the E^1 version of the Cauchy integral formula (Proposition 2.4.12) we have

$$g(z) = \frac{1}{2\pi i} \oint_{\partial \mathbb{D}_\pm} \frac{g(\xi)}{\xi - z} d\xi, \quad z \in \mathbb{D}_\pm.$$

Also notice that

$$g(z) = \frac{1}{2\pi i} \oint_{\partial \mathbb{D}_+} \frac{g(\xi)}{\xi - z} d\xi + \frac{1}{2\pi i} \oint_{\partial \mathbb{D}_-} \frac{g(\xi)}{\xi - z} d\xi, \quad z \in \mathbb{D}_+ \cup \mathbb{D}_-.$$

But since

$$\lim_{y\to 0^+} g(x+iy) = \lim_{y\to 0^-} g(x+iy)$$

almost everywhere on $[-1,1]$, the integrals over $[-1,1]$ cancel out and so

$$g(z) = \frac{1}{2\pi i}\oint_{\mathbb{T}} \frac{g(\xi)}{\xi - z}d\xi.$$

The above integral defines an analytic function on $\mathbb{D}$ and this proves the result. □

This next technical lemma is due to Smirnov [71] (see also [48, p. 319]).

Lemma 6.2.9. *For each $g \in N^+(\mathbb{D})$ there is a sequence $(g_n)_{n\geqslant 1} \subset H^\infty(\mathbb{D})$ such that*

1. *$|g_n(z)| \leqslant |g(z)|$ for all $z \in \mathbb{D}$;*

2. *$g_n \to g$ pointwise on $\mathbb{D}$.*

Proof. Factor g as $g = \theta h$, where θ is $\mathbb{D}$-inner and h is $\mathbb{D}$-outer [34, p. 74]. For each $n \in \mathbb{N}$ let h_n be the bounded $\mathbb{D}$-outer function whose boundary function satisfies

$$|h_n(\zeta)| = \begin{cases} |h(\zeta)|, & \text{if } |h(\zeta)| \leqslant n; \\ n, & \text{if } |h(\zeta)| > n \end{cases}$$

for almost every $\zeta \in \mathbb{T}$. We leave it to the reader to check, using properties of outer functions [34, p.73], that the functions $g_n := \theta h_n$, $n \in \mathbb{N}$, have the desired properties. □

Remark 6.2.10. Though not needed for what follows, we point out that if f is analytic on $\mathbb{D}$ and is a pointwise limit of a sequence of bounded analytic functions with increasing moduli, then $f \in N^+$ (see [48, p. 319] for a proof).

For an invariant subspace $\mathcal{M}$ of $H^2(G)$ and $\lambda_0 \in G \setminus Z(\mathcal{M})$, where $Z(\mathcal{M})$ is the set of common zeros for $\mathcal{M}$, let

$$\varphi := \frac{k_{\lambda_0}^{\mathcal{M}}}{\|k_{\lambda_0}^{\mathcal{M}}\|}$$

be the normalized reproducing kernel function for $\mathcal{M}$ (or equivalently the 'extremal function' for $\mathcal{M}$) at λ_0. Note that $\varphi \in \mathcal{M}$ and

$$\langle f, \varphi \rangle = \frac{f}{\varphi}(\lambda_0) \quad \forall f \in \mathcal{M}, \tag{6.2.11}$$

$$\langle \varphi, \varphi \rangle = 1.$$

These next two technical lemmas point out some special properties of this extremal function. But first we pause for a few remarks.

Remark 6.2.12.   1. Suppose $\alpha : G \rightarrow \widehat{\mathbb{C}} \setminus \gamma$ is the conformal map from (2.3.8) and $\mathcal{N} = C_{\alpha^{-1}} \mathcal{M}$. From the proof of Corollary 6.1.6 we know that $\mathcal{N}$ is nearly invariant. If

$$\Phi := \frac{k_0^{\mathcal{N}}}{\|k_0^{\mathcal{N}}\|}$$

is the normalized reproducing kernel function for $\mathcal{N}$ at 0, we can use the unitary operator $C_{\alpha^{-1}}$ to show that if we assume that $\lambda_0 = \alpha^{-1}(0)$, then

$$k_{\lambda_0}^{\mathcal{M}} = C_\alpha k_0^{\mathcal{N}}$$

and consequently

$$\varphi = \Phi \circ \alpha. \tag{6.2.13}$$

2. In what follows below, we need to be clear on how we represent the inner product in $H^2(G)$ as an integral. When Ω is Jordan domain with piecewise analytic boundary, the inner product in $H^2(\Omega)$ can be written as

$$\langle f, g \rangle = \int_{\partial \Omega} f \overline{g} \, d\omega_{\psi^{-1}(0), \Omega} = \int_{\partial \Omega} f \overline{g} |\psi'| \frac{ds}{2\pi},$$

where $\psi : \Omega \rightarrow \mathbb{D}$. For the slit domain $G = \mathbb{D} \setminus [0, 1)$ the expression

$$\int_{\partial G} f \overline{g} d\omega,$$

is not quite right since we need to take into account the fact that $f^+ \overline{g^+}$ and $f^- \overline{g^-}$ are, in general, different. Thus we will use the notation

$$\int_{\partial G} f \overline{g} d\omega^*$$

to mean

$$\int_{\mathbb{T}} f(\zeta) \overline{g(\zeta)} w(\zeta) \frac{|d\zeta|}{2\pi} + \int_0^1 (f^+(x) \overline{g^+(x)} + f^-(x) \overline{g^-(x)}) w(x) \frac{dx}{2\pi},$$

where $w = |\psi'|$. Note that we really should have w^+ and w^- in the above expression. However, recall from (2.3.2) that $w^+ = w^-$. Also observe from Proposition 2.3.4 that this last expression is precisely $\langle f, g \rangle$, the inner product in $H^2(G)$.

Lemma 6.2.14. *If the greatest common G-inner divisor of an invariant subspace $\mathcal{M}$ of $H^2(G)$ is equal to 1, then the normalized reproducing kernel function φ for $\mathcal{M}$ at $\lambda_0 \in G \setminus Z(\mathcal{M})$ extends analytically across $\mathbb{T} \setminus \{1\}$.*

Proof. For each $|\lambda| > 1$, note that

$$\frac{z - \lambda_0}{z - \lambda} \in H^\infty(\mathbb{D})$$

and so, by Proposition 6.1.1,

$$\frac{z - \lambda_0}{z - \lambda} f \in \mathcal{M} \quad \forall f \in \mathcal{M} \setminus \{0\}.$$

Thus, by the reproducing property of φ at λ_0 (see (6.2.11)), we have

$$\int_{\partial G} \frac{z - \lambda_0}{z - \lambda} f(z) \overline{\varphi}(z) d\omega^*(z) = \left\langle \frac{z - \lambda_0}{z - \lambda} f, \varphi \right\rangle = 0. \tag{6.2.15}$$

Take note of Remark 6.2.12.

Combining Fatou's jump theorem (Theorem 3.3.3) and (6.2.15), we get

$$\lim_{r \to 1^-} \int_{\partial G} \frac{z - \lambda_0}{z - r\zeta} f(z) \overline{\varphi}(z) d\omega^*(z) = \overline{\zeta}(\zeta - \lambda_0) f(\zeta) \overline{\varphi}(\zeta) w(\zeta) \quad \text{a.e. } \zeta \in \mathbb{T}. \tag{6.2.16}$$

Notice how the contribution from the integrals over the slit cancels out in the limit.

From Remark 6.2.3 (in particular (6.2.4)) $w(\zeta)$ is equal to $|\psi'(\zeta)|$ and from elementary facts about conformal mappings, ψ' is analytically continuable across $\mathbb{T} \setminus \{1\}$ and the analytic continuation has no zeros in an open neighborhood of $\mathbb{T} \setminus \{1\}$.[3] It follows, for some appropriate branch of $\sqrt{\cdot}$, that the function

$$W(z) := \sqrt{\psi'(z)}$$

has the same property. Note that $w(\zeta) = W(\zeta)\overline{W(\zeta)}$ for $\zeta \in \mathbb{T} \setminus \{1\}$.

This means that the identity in (6.2.16) can be re-written as

$$\overline{\zeta}\,\overline{\varphi}(\zeta)\overline{W}(\zeta) = \frac{1}{(\zeta - \lambda_0)f(\zeta)W(\zeta)} \lim_{r \to 1^-} \int_{\partial G} \frac{z - \lambda_0}{z - r\zeta} f(z) \overline{\varphi}(z) d\omega^*(z) \quad \text{a.e. } \zeta \in \mathbb{T}. \tag{6.2.17}$$

Now select a $\zeta_0 \in \mathbb{T} \setminus \{1\}$ and an open disk $\Delta := \{|z - \zeta_0| < r\}$ contained in the region of analyticity of W and such that $\lambda_0 \notin \Delta$. Consider the function F on $\Delta \cap \mathbb{D}_e$ defined by

$$F(\lambda) := \frac{1}{\lambda} \overline{\varphi}\left(\frac{1}{\overline{\lambda}}\right) \overline{W}\left(\frac{1}{\overline{\lambda}}\right).$$

Observe that F is analytic on $\Delta \cap \mathbb{D}_e$, $F|\mathbb{T}$ is almost everywhere equal to the left-hand side of (6.2.17), and, as a consequence, is integrable on $\Delta \cap \mathbb{T}$. (Note that φ is integrable on $\Delta \cap \mathbb{T}$ and W is bounded on $\mathbb{T}$.) Adjusting the radius of Δ slightly, we can assume that F is also bounded on $(\partial\Delta) \cap \mathbb{D}_e$. But since $F \in N^+(\Delta \cap \mathbb{D}_e)$ and has integrable boundary values, we see from Proposition 2.4.10 that $F \in E^1(\Delta \cap \mathbb{D}_e)$.

Now consider the function F_1 on $\Delta \cap \mathbb{D}$ defined by

$$F_1(\lambda) := \frac{1}{(\lambda - \lambda_0)f(\lambda)W(\lambda)} \int_{\partial G} \frac{z - \lambda_0}{z - \lambda} f(z) \overline{\varphi}(z) d\omega^*(z). \tag{6.2.18}$$

---

[3]One can also see this by looking at the exact form of ψ which one can compute from the appendix.

Since the greatest common G-inner divisor of $\mathcal{M}$ is equal to 1, we conclude that $F_1 \in N^+(\Delta \cap \mathbb{D})$ [4] and, adjusting the radius of Δ, is bounded on $(\partial \Delta) \cap \mathbb{D}$. Note also from (6.2.17) and the discussion in the previous paragraph, that F_1 is integrable on $\Delta \cap \mathbb{T}$. Thus $F_1 \in E^1(\Delta \cap \mathbb{D})$.

Finally, $F(\zeta) = F_1(\zeta)$ for almost every $\zeta \in \Delta \cap \mathbb{T}$ and so, by Proposition 6.2.8 (Morera's theorem), F_1 is an analytic continuation of F across $\Delta \cap \mathbb{T}$. Since F has an analytic continuation to Δ and W is analytic on Δ, one can look at the formula defining F to see that φ has an analytic continuation across $\Delta \cap \mathbb{T}$. The lemma now follows. $\qquad\square$

Corollary 6.2.19. *The normalized reproducing kernel function φ in Lemma 6.2.14 has no zeros on $[-1,0)$.*

The proof of Corollary 6.2.19 requires some information about the boundary values of the $\mathbb{D}_+$-outer function F in the statement of Corollary 6.1.6. Recall from the proof of Corollary 6.1.6 that

$$F = F_1 \circ (\alpha|\mathbb{D}_+),$$

where $\alpha : G \to \widehat{\mathbb{C}} \setminus \gamma$ is from (2.3.8) and F_1 is the $\mathbb{D}$-outer function from the proof of Theorem 3.1.2 (see (3.5.1)). More precisely, let Φ be the normalized reproducing kernel function at the origin for the nearly invariant subspace $C_{\alpha^{-1}}\mathcal{M}$ and let F_1 be the $\mathbb{D}$-outer function whose non-tangential boundary function satisfies

$$|F_1|^2 = \frac{|\Phi_i|^2}{1 + w_\gamma \chi_\gamma |\Phi_e|^2} \quad \text{a.e on } \mathbb{T},$$

where

$$\Phi_i(\zeta) = \lim_{r \to 1^-} \Phi(r\zeta), \quad \Phi_e(\zeta) = \lim_{r \to 1^+} \Phi(r\zeta) \quad \text{a.e. } \zeta \in \mathbb{T}.$$

Let us now compute the non-tangential boundary function for $|F|$. For $\zeta \in \mathbb{T}$ notice that

$$|F_1(\zeta)|^2 = \angle \lim_{z \to \zeta} \frac{|\Phi(z)|^2}{1 + w_\gamma(\zeta)\chi_\gamma(\zeta)|\Phi(\frac{1}{\bar{z}})|^2},$$

where $\angle \lim$ denotes the non-tangential limit as $z \to \zeta, z \in \mathbb{D}$. Thus for almost every $\xi \in \mathbb{T}$ with $0 < \arg(\xi) < \pi$ we have

$$|F(\xi)|^2 = |F_1 \circ \alpha(\xi)|^2$$
$$= \lim_{r \to 1^-} \frac{|\Phi \circ \alpha(r\xi)|^2}{1 + (w_\gamma \circ \alpha)(\xi)|\Phi(\frac{1}{\overline{\alpha(r\xi)}})|^2} \quad (\text{since } \alpha(\mathbb{D}_+) = \mathbb{D})$$
$$= \lim_{r \to 1^-} \frac{|\Phi \circ \alpha(r\xi)|^2}{1 + w_G(\xi)|\Phi(\frac{1}{\overline{\alpha(r\xi)}})|^2} \quad (\text{since } w_\gamma \circ \alpha = w_G)$$

---

[4]By (6.2.16), the non-tangential boundary values of F_1 are the same for every $f \in \mathcal{M} \setminus \{0\}$. By Privalov's uniqueness theorem [51, p. 62], the definition of F_1 is independent of f. Using the fact that the greatest common divisor of $\mathcal{M}$ is 1, it can be argued, by adjusting f, that no part of an G-inner factor may appear in the denominator of the definition of F_1.

$$= \lim_{r \to 1^-} \frac{|\Phi \circ \alpha(r\xi)|^2}{1 + w_G(\xi)|\Phi \circ \alpha(r\overline{\xi})|^2} \quad \text{(since } 1/\overline{\alpha(r\xi)} = \alpha(r\overline{\xi}))$$

$$= \frac{|\Phi \circ \alpha(\xi)|^2}{1 + w_G(\xi)|\Phi \circ \alpha(\overline{\xi})|^2}$$

$$= \frac{|\varphi(\xi)|^2}{1 + w_G(\xi)|\varphi(\overline{\xi})|^2} \quad \text{(since } \varphi = \Phi \circ \alpha \text{ from (6.2.13)).}$$

For almost every $x \in [0,1]$ we have

$$|F(x)|^2 = \lim_{y \to 0^+} |F_1 \circ \alpha(x+iy)|^2$$

$$= \lim_{y \to 0^+} \frac{|\Phi \circ \alpha(x+iy)|^2}{1 + w_G(x)|\Phi(\frac{1}{\alpha(x+iy)})|^2}$$

$$= \lim_{y \to 0^+} \frac{|\Phi \circ \alpha(x+iy)|^2}{1 + w_G(x)|\Phi \circ \alpha(x-iy)|^2} \quad \text{(since } 1/\overline{\alpha(z)} = \alpha(\overline{z}))$$

$$= \frac{|\varphi^+(x)|^2}{1 + w_G(x)|\varphi^-(x)|^2}$$

since $\alpha(x+iy) \in \mathbb{D}, \alpha(x-iy) \in \mathbb{D}_e, \alpha^+(x) = \alpha^-(x), \alpha([0,1]) = \{e^{i\theta} : \frac{3\pi}{2} \leqslant \theta \leqslant 2\pi\}$. For $x \in (-1,0)$ we have $|F(x)|^2 = |\varphi(x)|^2$.

To summarize, F is the $\mathbb{D}_+$-outer function whose boundary function (almost everywhere) satisfies

$$|F(\xi)|^2 = \begin{cases} \dfrac{|\varphi(\xi)|^2}{1 + w_G(\xi)|\varphi(\overline{\xi})|^2}, & \xi \in \mathbb{T}_+; \\ |\varphi(\xi)|^2, & \xi \in [-1,0]; \\ \dfrac{|\varphi^+(\xi)|^2}{1 + w_G(\xi)|\varphi^-(\xi)|^2}, & \xi \in [0,1]. \end{cases} \tag{6.2.20}$$

In the above definitions, and for what follows, we will use the notation

$$\mathbb{T}_+ := \{e^{i\theta} : 0 \leqslant \theta \leqslant \pi\} \quad \text{and} \quad \mathbb{T}_- := \{e^{i\theta} : \pi \leqslant \theta \leqslant 2\pi\}.$$

Proof of Corollary 6.2.19. Let $\lambda \in (-1,0)$ and let I_λ be a closed sub-interval of $(-1,0)$ that contains λ in its interior. From the formula for F in (6.2.20), notice that

$$|F| = |\varphi| \quad \text{a.e. on } (-1,0). \tag{6.2.21}$$

Notice also from elementary facts about conformal maps and (6.2.2) that

$$w_+|I_\lambda \text{ is bounded above and below.} \tag{6.2.22}$$

Since $\lambda \in G$ and $\Theta_{\mathcal{M}} \equiv 1$, there must be an $f \in \mathcal{M}$ such that

$$f \text{ is never zero on } I_\lambda. \tag{6.2.23}$$

But since $f|\mathbb{D}_+/F \in H^2(\mathbb{D}_+)$ we have, from applying (6.2.22) followed by (6.2.23) followed by (6.2.21),

$$\infty > \int_{I_\lambda} \left|\frac{f}{F}\right|^2 w_+ dx \geqslant c \int_{I_\lambda} \left|\frac{f}{F}\right|^2 dx \geqslant c \int_{I_\lambda} \frac{1}{|F|^2} dx = c \int_{I_\lambda} \frac{1}{|\varphi|^2} dx.$$

Since φ is analytic in a neighborhood of I_λ, the only way that

$$\int_{I_\lambda} \frac{1}{|\varphi|^2} dx < \infty$$

is for φ to have no zeros on I_λ. Thus we have shown that φ has no zeros on $(-1,0)$.

We will now argue that $\varphi(-1) \neq 0$. Let J be the arc of the unit circle subtended by the points -1 and i. Let $f \in \mathcal{M} \setminus \{0\}$ and let f_1 be the $\mathbb{D}$-outer function whose boundary values satisfy[5]

$$|f_1| := \begin{cases} \dfrac{1}{|f|}, & \text{a.e. on } J; \\ 1, & \text{a.e. on } \mathbb{T} \setminus J. \end{cases} \tag{6.2.24}$$

By the definition of $\mathbb{D}$-outer we have

$$f_1(z) = \exp\left(-\int_J \frac{\zeta+z}{\zeta-z} \log|f(\zeta)| dm(\zeta)\right)$$

and so, since the integration is over J, f_1 is bounded on $[0,1)$. Clearly we have $ff_1 \in N^+(G)$ [6] and $f_1 f \in L^2(\partial G, \omega)$. By Proposition 2.4.10, $f_1 f \in H^2(G)$. Use Lemma 6.2.9 to produce a sequence $(g_n)_{n \geqslant 1}$ in $H^\infty(\mathbb{D})$ with $g_n \to f_1$ pointwise in $\mathbb{D}$ and $|g_n| \leqslant |f_1|$ on $\mathbb{D}$. By Proposition 6.1.1, $g_n f \in \mathcal{M}$ for each n. Moreover, $g_n f \to f_1 f$ pointwise in G. We also see that $|g_n f|^2 \leqslant |f_1 f|^2$ on G and so, by the harmonic majorant definition of the norm on $H^2(G)$, $\|g_n f\|_{H^2(G)}$ is uniformly bounded in n. Thus $g_n f \to f_1 f$ weakly and so $f_1 f \in \mathcal{M}$.

Moreover, from (6.2.2) and (6.2.5),

$$w_+|J \asymp |z+1|. \tag{6.2.25}$$

Since $f_1 f \in \mathcal{M}$, then $f_1 f|\mathbb{D}_+/F \in H^2(\mathbb{D}_+)$ and so, by using (6.2.24) followed by (6.2.20) followed by (6.2.25),

$$\infty > \int_J \left|\frac{f_1 f}{F}\right|^2 w_+ |d\zeta| \geqslant \int_J \frac{w_+}{|F|^2} |d\zeta| \geqslant \int_J \frac{w_+}{|\varphi|^2} |d\zeta| \geqslant c \int_J \frac{|\zeta+1|}{|\varphi(\zeta)|^2} |d\zeta|.$$

Since φ is analytic in a neighborhood of -1 (Lemma 6.2.14), the only way this last integral can be finite is for $\varphi(-1) \neq 0$. $\qquad\square$

---

[5] By (2.4.3), $\log|f| \in L^1(\partial G, \omega)$ and from (2.3.6) $d\omega \asymp d\theta$ on J. Thus it follows that $\log|f_1| \in L^1(\mathbb{T}, m)$ and so such a $\mathbb{D}$-outer function actually exists.

[6] Observe that $f \in H^2(G) \subset N^+(G)$ and f_1 is $\mathbb{D}$-outer and hence, by Proposition 2.4.5, f_1 is G-outer.

Lemma 6.2.26. *For each $\varepsilon \in (0,1)$, the $\mathbb{D}_+$-outer function from (6.2.20) satisfies*

$$\int_{\mathbb{T}_+ \cup [-\varepsilon,1]} |\log |F|| w_\varepsilon ds < \infty.$$

Proof. Using (6.2.20), write

$$|F|^2 = \frac{|\psi_1|^2}{1 + w_G |\psi_2|^2}, \quad \text{a.e. on } \mathbb{T}_+ \cup [0,1].$$

We see that

$$\int_{\mathbb{T}_+ \cup [0,1]} |\log |F|^2| w_\varepsilon ds \leqslant \int_{\mathbb{T}_+ \cup [0,1]} |\log |F|^2| w ds \quad \text{(by Lemma 6.2.6)}$$

$$\leqslant \int_{\mathbb{T}_+ \cup [0,1]} |\log |\psi_1|^2| w ds + \int_{\mathbb{T}_+ \cup [0,1]} \log(1 + w|\psi_2|^2) w ds.$$

The first integral in the previous line converges since ψ_1 is part of the boundary function for φ and $\varphi \in H^2(G) \setminus \{0\}$ (see (2.4.3)). For the second integral, use the inequality

$$\log(1 + y) \leqslant 1 + |\log y|, \quad y > 0,$$

to show that this integral is bounded above by

$$\int_{\mathbb{T}_+ \cup [0,1]} w ds + \int_{\mathbb{T}_+ \cup [0,1]} |\log w| w ds + \int_{\mathbb{T}_+ \cup [0,1]} |\log |\psi_2|^2| w ds.$$

The first integral clearly converges. The second integral converges by Lemma 6.2.7(1) while the third integral converges since ψ_2 is part of the boundary function for φ and $\varphi \in H^2(G)$.

We are now left with showing that the integral

$$\int_{[-\varepsilon,0]} |\log |F|| w_\varepsilon ds$$

converges. But this one is easy since, by (6.2.20), $|F| = |\varphi|$ a.e. on $[-\varepsilon,0]$ and $\varphi \in H^2(G_\varepsilon) \setminus \{0\}$. (This last fact follows from the fact that $\varphi \in H^2(G)$ and $G_\varepsilon \subset G$ – see (2.1.4)). This completes the proof. $\square$

With these technical details out of the way, we are finally ready for the proof of Theorem 6.2.1.

Proof of Theorem 6.2.1. Let $\varepsilon \in (0,1)$. We leave it to the reader to use Lemma 6.2.7 and Lemma 6.2.26 to verify that there is a G_ε-outer function F_ε whose boundary function satisfies

$$|(F_\varepsilon)^+|^2 = \frac{w_\varepsilon}{w_+} |F^+|^2 \text{ a.e. on } [-\varepsilon, 1];$$

$$|F_\varepsilon|^2 = \frac{w_\varepsilon}{w_+} |F|^2 \text{ a.e. on } \mathbb{T}_+;$$

$$|F_\varepsilon|^2 = 1 \quad \text{a.e. on } \mathbb{T}_-;$$
$$|(F_\varepsilon)^-|^2 = 1 \quad \text{a.e. on } [-\varepsilon, 1].$$

To finish the proof, we need to show that for $f \in H^2(G)$

$$\frac{f|\mathbb{D}_+}{F} \in H^2(\mathbb{D}_+) \Leftrightarrow \frac{f}{F_\varepsilon} \in H^2(G_\varepsilon),$$

or equivalently that

$$I_1 := \int_{\partial\mathbb{D}_+} \left|\frac{f}{F}\right|^2 w_+ ds < \infty$$

if and only if

$$I_2 := \int_{\mathbb{T}} \left|\frac{f}{F_\varepsilon}\right|^2 w_\varepsilon ds + \int_{[-\varepsilon,1]} \left(\left|\frac{f^+}{(F_\varepsilon)^+}\right|^2 + \left|\frac{f^-}{(F_\varepsilon)^-}\right|^2\right) w_\varepsilon ds < \infty.$$

By the construction of F_ε above we have

$$I_2 \leqslant I_1 + \int_{[-\varepsilon,1]} |f^-|^2 w_\varepsilon ds + \int_{\mathbb{T}_-} |f|^2 w_\varepsilon ds$$
$$\leqslant I_1 + c\|f\|^2_{H^2(G)} \quad \text{(by Lemma 6.2.6)}$$

and, since $|F|^2 = |\varphi|^2$ on $[-1, -\varepsilon]$ and φ is non-zero on $[-1, -\varepsilon]$ (Corollary 6.2.19),

$$I_1 \leqslant I_2 + \int_{[-1,-\varepsilon]} \frac{|f|^2}{|F|^2} w_+ ds$$
$$\leqslant I_2 + c\|f\|^2_{H^2(\mathbb{D}_+)}$$
$$\leqslant \qquad\qquad I_2 + c\|f\|^2_{H^2(G)} \quad \text{(by (2.1.4))}.$$

$\square$

Remark 6.2.27. As one can see from the very end of the proof of Theorem 6.2.1, the fact that $\varepsilon > 0$ is important since the constant c in the last two lines of the proof depends on ε. We do not know whether or not the condition '$f/F_\varepsilon \in H^2(G_\varepsilon)$' (where F_ε is some G_ε-outer function) can be replaced by '$f/F \in H^2(G)$' (where F is some G-outer function).

Chapter 7

Cyclic invariant subspaces

7.1 Two-cyclic subspaces

If $\mathcal{M}$ is an invariant subspace of $H^2(G)$, the proof of Corollary 6.1.6 shows that $\mathcal{N} := C_{\alpha^{-1}} \circ \mathcal{M}$ is a nearly invariant subspace of $H^2(\widehat{\mathbb{C}} \setminus \gamma)$. We know from Corollary 3.2.9 that if $\{0, \infty\}$ is not a subset of the common zeros of $\mathcal{N}$ and Φ and Ψ are the normalized reproducing kernels at $z = 0$ and $z = \infty$, then the smallest nearly invariant subspace containing Φ and Ψ is equal to $\mathcal{N}$. From Remark 6.2.12 we also see that $\Phi \circ \alpha$ is the normalized reproducing kernel for $\mathcal{M}$ at $\alpha^{-1}(0)$ while $\Psi \circ \alpha$ is the normalized reproducing kernel at $\alpha^{-1}(\infty)$.

Theorem 7.1.1. *If $\mathcal{M}$ is a non-trivial invariant subspace of $H^2(G)$, then*

$$\mathcal{M} = \bigvee \left\{ z^n (\Phi \circ \alpha), z^m (\Psi \circ \alpha) : n, m \in \mathbb{N}_0 \right\}.$$

Proof. Without loss of generality, we assume that $\alpha^{-1}(0)$ and $\alpha^{-1}(\infty)$ do not belong to the common zero set of $\mathcal{M}$. We know that $C_{\alpha^{-1}} \mathcal{M} = \mathcal{N}_{\Phi, \Psi}$, where $\mathcal{N}_{\Phi, \Psi}$ is the smallest nearly invariant subspace containing Φ and Ψ. Thus we have

$$\mathcal{M}_{\Phi \circ \alpha, \Psi \circ \alpha} \subset \mathcal{M} = C_{\alpha} \mathcal{N}_{\Phi, \Psi},$$

where $\mathcal{M}_{\Phi \circ \alpha, \Psi \circ \alpha}$ is the smallest invariant subspace of $H^2(G)$ containing $\Phi \circ \alpha$ and $\Psi \circ \alpha$. This means that $C_{\alpha^{-1}} \mathcal{M}_{\Phi \circ \alpha, \Psi \circ \alpha}$ is a nearly invariant subspace of $H^2(\widehat{\mathbb{C}} \setminus \gamma)$ which contains Φ and Ψ and so, by definition, $C_{\alpha^{-1}} \mathcal{M}_{\Phi \circ \alpha, \Psi \circ \alpha} = \mathcal{N}_{\Phi, \Psi}$. The result now follows. $\qquad \square$

For general functions $f, g \in H^2(G)$, when is the invariant subspace generated by f and g equal to all of $H^2(G)$?

Theorem 7.1.2. *If $f, g \in H^2(G) \setminus \{0\}$, then*

$$\bigvee \{ z^n f, z^m g : n, m \in \mathbb{N}_0 \} = H^2(G).$$

if and only if f and g have no non-trivial common G-inner factor and the set

$$\left\{ x \in [0,1) : \frac{f^+(x)}{f^-(x)} = \frac{g^+(x)}{g^-(x)} \right\}$$

has Lebesgue measure zero.

Proof. One direction is easy. For the other direction, let $\mathcal{M}$ be the invariant subspace generated by f and g. By Corollary 6.1.6,

$$\mathcal{M} = \Theta \cdot \left\{ h \in H^2(G) : \frac{h|\mathbb{D}_+}{F} \in H^2(\mathbb{D}_+), h^+ = \rho h^- \text{ a.e. on } E \right\}.$$

Since f and g have no common G-inner factor, it must be the case that $\Theta \equiv 1$. Also, since the functions f^+/f^- and g^+/g^- are equal almost nowhere, E has measure zero. Thus

$$\mathcal{M} = \left\{ h \in H^2(G) : \frac{f|\mathbb{D}_+}{F} \in H^2(\mathbb{D}_+) \right\}.$$

This means that $\mathcal{M}$ is $H^\infty(G)$-invariant and so, by Proposition 6.1.2, $\mathcal{M} = \Theta_1 H^2(G)$ for some G-inner function Θ_1. But again, since f and g have no common G-inner factor, we must have $\Theta_1 \equiv 1$ and so $\mathcal{M} = H^2(G)$. $\square$

Example 7.1.3. The invariant subspace generated by the functions 1 and $\sqrt{z}$ is $H^2(G)$.

7.2   Cyclic subspaces

Theorem 7.1.1 says that every invariant subspace $\mathcal{M}$ is *2-cyclic* in the sense that it is generated by two functions. Do we really need *both* functions to generate $\mathcal{M}$? Is $\mathcal{M}$ always *cyclic*, i.e., is there a single $f \in \mathcal{M}$ so that

$$\mathcal{M} = [f] := \bigvee \{ z^n f : n \in \mathbb{N}_0 \}?$$

When $\mathcal{M} = H^2(G)$, results from [4][1] show that $\mathcal{M}$ is not cyclic (see also Remark 7.2.1 below). What are the cyclic invariant subspaces of $H^2(G)$?

Remark 7.2.1. It is easy to see that

$$[f] \subset \mathcal{M}(\rho) := \{ h \in H^2(G) : h^+ = \rho h^- \text{ a.e. on } [0,1] \},$$

where $\rho = f^+/f^-$ almost everywhere. Suppose that an invariant subspace $\mathcal{M} \neq \{0\}$, with parameters $\Theta, F_\varepsilon (0 < \varepsilon < 1), \rho, E$ from Theorem 6.2.1 (note that Θ, ρ, and E are essentially unique – Corollary 3.6.3) is cyclic. Then $m_1([0,1] \setminus E) = 0$. Here m_1 is Lebesgue measure on $[0,1]$. Indeed, suppose $m_1([0,1] \setminus E) > 0$. Then there is a closed subset F of $[0,1] \setminus E$ with $m_1(F) > 0$. Using (3.6.1), one produces a $g \in H^\infty(\mathbb{D} \setminus F) \setminus \{0\}$ with

$$\frac{g^+}{g^-} \neq 1$$

---

[1]The follow-up papers [2, 3] discuss other cyclicity problems.

on some compact subset of F of positive measure. From the definition of $\mathcal{M}$ we see that $g\mathcal{M} \subset \mathcal{M}$. Thus if $[f] = \mathcal{M}$, then $gf \in \mathcal{M}$. However $\mathcal{M} = [f] \subset \mathcal{M}(\rho)$ as above and so $(gf)^+/(gf)^- = \rho$ almost everywhere on $[0,1]$. But g was constructed so that this last equality can not hold almost everywhere on F. Thus

$$\mathcal{M} \text{ is cyclic} \Rightarrow m_1([0,1] \setminus E) = 0.$$

We will see in Example 8.2.13, using an analysis of the essential spectrum of $S|\mathcal{M}$, that the other direction does not hold.

The next few results compute $[f]$ for certain reasonably well-behaved $f \in H^2(G)$.

Theorem 7.2.2. *Suppose both h and $1/h$ belong to $H^2(G)$. Then $[h] = \mathcal{M}(\rho)$, where $\rho = h^+/h^-$.*

Proof. So far we have $[h] \subset \mathcal{M}(\rho)$. To see the other direction, suppose $f \in \mathcal{M}(\rho)$. Using the fact that $1/h \in H^2(G)$ and the Cauchy-Schwarz inequality, we see that $g := f/h \in H^1(G)$ and so by (2.1.4),

$$g|\mathbb{D}_+ \in H^1(\mathbb{D}_+) \quad \text{and} \quad g|\mathbb{D}_- \in H^1(\mathbb{D}_-). \tag{7.2.3}$$

Let $q := w_3 \circ w_2 \circ w_1$ be a conformal map from $\mathbb{D}$ onto $\mathbb{D}_+$ (see the appendix). A computation shows that the function $(1+q)(1-q)q'$ is bounded. From the conformal invariance of the Hardy spaces and (7.2.3), we have that $g|\mathbb{D}_+ \circ q \in H^1(\mathbb{D})$ and consequently, the function $g_1(z) := (1+z)(1-z)g(z)$ has the property that $g_1|\mathbb{D}_+ \in E^1(\mathbb{D}_+)$, i.e., $(g_1|\mathbb{D}_+ \circ q)q' \in H^1(\mathbb{D})$ – see (2.4.9). In a similar way, $g_1|\mathbb{D}_- \in E^1(\mathbb{D}_-)$.

Using the Cauchy integral formula (Proposition 2.4.12) we have

$$g_1(z) = \frac{1}{2\pi i} \oint_{\partial \mathbb{D}_\pm} \frac{g_1(\zeta)}{\zeta - z} d\zeta, \quad z \in \mathbb{D}_\pm.$$

However, $(g_1)^+ = (g_1)^-$ almost everywhere on $[-1,1]$ and so for all $z \in \mathbb{D} \setminus [-1,1]$ we have

$$g_1(z) = \frac{1}{2\pi i} \oint_{\partial \mathbb{D}_+} \frac{g_1(\zeta)}{\zeta - z} d\zeta + \frac{1}{2\pi i} \oint_{\partial \mathbb{D}_-} \frac{g_1(\zeta)}{\zeta - z} d\zeta = \frac{1}{2\pi i} \oint_{\mathbb{T}} \frac{g_1(\zeta)}{\zeta - z} d\zeta.$$

Notice in the above calculation how the integrals on $[-1,1]$ cancel each other out. This means that g_1 is a Cauchy transform of a measure on $\mathbb{T}$ and consequently g_1 has an analytic continuation across $[-1,1]$ to a function which belongs to $H^p(\mathbb{D})$ for all $0 < p < 1$ [31, p. 39]. In particular, $g \in N^+(\mathbb{D})$.

By Lemma 6.2.9, there is a sequence $(g_n)_{n \geqslant 1}$ in $H^\infty(\mathbb{D})$ such that $g_n \to g$ pointwise in $\mathbb{D}$ as $n \to \infty$ and $|g_n| \leqslant |g|$ on $\mathbb{D}$. Hence $g_n h \to f$ pointwise in G and $|g_n h|^2 \leqslant |gh|^2 = |f|^2$ on G. This last inequality says that the $H^2(G)$ norms of $g_n h$ are uniformly bounded. Thus $g_n h \in [h]$ (Proposition 6.1.1) and $g_n h \to f$ weakly in $H^2(G)$ which means that $f \in [h]$. $\square$

It is routine to show that if f and $1/f$ belong to $H^2(G)$, then f is G-outer [34, p. 68]. One might conjecture that if $f \in H^2(G)$ is G-outer, then

$$[f] = \mathcal{M}(\rho),$$

where $\rho = f^+/f^-$. However this is not the case.

Example 7.2.4. Consider the *G*-outer function $f(z) = z$. By (7.3.1) and Theorem 7.3.2 (see below)

$$[z] = \{g \in \mathcal{M}(1) : g(0) = 0\}$$

which is a proper subset of $\mathcal{M}(1)$.

Remark 7.2.5. One might wonder where the classical Hardy space $H^2(\mathbb{D})$ fits in with $\mathcal{M}(1)$. They do look very similar. It follows from (2.1.4) that $H^2(\mathbb{D}) \subset \mathcal{M}(1)$ with continuous embedding. However, this containment is proper. For example, the function

$$f(z) = \frac{1}{\sqrt{1-z}}$$

is analytic across $[0,1]$ but does not belong to $H^2(\mathbb{D})$ since

$$\|f\|_{H^2(\mathbb{D})}^2 = \int_0^{2\pi} \frac{1}{|1-e^{i\theta}|} \frac{d\theta}{2\pi} = \infty.$$

However, $f \in N^+(G)$ and by (6.2.5) $f|\partial G \in L^2(\partial G, \omega^*)$. Thus, by Proposition 2.4.10, $f \in H^2(G)$ and hence, since f is analytic on $\mathbb{D}$, $f \in \mathcal{M}(1)$.

Corollary 7.2.6. *Suppose $f = \Theta f_1$, where Θ is G-inner and f_1 is G-outer such that f_1 and $1/f_1$ belong to $H^2(G)$. Then*

$$[f] = \Theta \cdot \mathcal{M}(\rho),$$

where $\rho = f_1^+ / f_1^-$.

Proof. Use the fact that Θ is *G*-inner and so multiplication by Θ is an isometry to argue that $[f] = \Theta \cdot [f_1]$. Now use Theorem 7.2.2. $\square$

7.3   Polynomial approximation

Our results have applications to polynomial approximation and analytic bounded point evaluations. Let ω be the harmonic measure for ∂G and $P^2(\omega)$ be the closure of the analytic polynomials in $L^2(\omega)$. Notice that

$$[1] = P^2(\omega) \subset \mathcal{M}(1).$$

Since $d\omega|\mathbb{T} \asymp |\psi'|dm$, where ψ is the conformal map from G onto $\mathbb{D}$, and since $\theta \to \psi'(e^{i\theta})$ is a log-integrable bounded function on $[0, 2\pi]$ (see (6.2.5)), there is a bounded $\mathbb{D}$-outer function F on $\mathbb{D}$ such that $|\psi'| = |F|^2$ almost everywhere on $\mathbb{T}$. For an analytic polynomial p we can apply the Cauchy integral formula (see (2.4.11)) to see that for fixed $a \in \mathbb{D}$,

$$p(a)F(a) = \int_{\mathbb{T}} \frac{p(\zeta)F(\zeta)}{1 - \bar{\zeta}a} dm(\zeta).$$

The Cauchy-Schwarz inequality yields

$$|p(a)F(a)| \leqslant C_a \left(\int_{\mathbb{T}} |p|^2 |F|^2 dm \right)^{1/2}$$

$$\leqslant C_a \left(\int_{\mathbb{T}} |p|^2 d\omega \right)^{1/2}$$

$$\leqslant C_a \|p\|_{L^2(\omega)}.$$

Divide through by $F(a)$ (which is never zero since F is $\mathbb{D}$-outer) to get

$$|p(a)| \leqslant c_a \|p\|_{L^2(\omega)} \tag{7.3.1}$$

for all polynomials p. In other words, $\mathbb{D}$ is the set of *bounded point evaluations* for $P^2(\omega)$. This also means that the linear functional $p \mapsto p(a)$, initially defined on the analytic polynomials, continues to a bounded linear functional on $P^2(\omega)$.

Theorem 7.3.2. *For $f \in H^2(G)$, the following are equivalent.*

1. *$f \in P^2(\omega)$;*

2. *$f \in \mathcal{M}(1)$;*

3. *f has an analytic continuation to $\mathbb{D}$;*

4. *$f \in \mathrm{clos}_{H^2(G)} H^\infty(\mathbb{D})$.*

Proof. The equality $[1] = P^2(\omega)$ is clear. Theorem 7.2.2 gives us $[1] = \mathcal{M}(1)$ and so (1) $\Leftrightarrow$ (2). From the proof of Proposition 6.1.1 we have

$$[1] = \mathrm{clos}_{H^2(G)} H^\infty(\mathbb{D})$$

and so (1) $\Leftrightarrow$ (4). Now use the conformal map $\alpha : G \to \widehat{\mathbb{C}} \setminus \gamma$ along with Corollary 4.2.20 to see that (3) $\Leftrightarrow$ (4). $\qquad \square$

Chapter 8

The essential spectrum

8.1 Fredholm theory

If $\mathcal{B}(\mathcal{H})$ is the algebra of bounded linear operators on a Hilbert space $\mathcal{H}$ and $\mathcal{K}$ is the ideal of compact operators on $\mathcal{H}$, one forms the *Calkin algebra* $\mathcal{B}(\mathcal{H})/\mathcal{K}$ and the natural map $\pi : \mathcal{B}(\mathcal{H}) \to \mathcal{B}(\mathcal{H})/\mathcal{K}$. Recall that $A \in \mathcal{B}(\mathcal{H})$ is *Fredholm* if $\pi(A)$ is invertible in $\mathcal{B}(\mathcal{H})/\mathcal{K}$. A well-known theorem [19, p. 356] says that A is Fredholm precisely when $\operatorname{Rng} A$ is closed and both $\ker A$ and $\mathcal{H}/\operatorname{Rng} A$ are finite dimensional. An operator A is *semi-Fredholm* if $\pi(A)$ is either right or left invertible in $\mathcal{B}(\mathcal{H})/\mathcal{K}$. Equivalently, A is semi-Fredholm if and only if $\operatorname{Rng} A$ is closed and either $\ker(A)$ or $\mathcal{H}/\operatorname{Rng} A$ is finite dimensional. We also use the notation

$$\sigma(A) := \{\lambda \in \mathbb{C} : \lambda I - A \text{ is not invertible}\} \quad \text{(spectrum of } A\text{)},$$

$$\sigma_e(A) := \{\lambda \in \mathbb{C} : \lambda I - A \text{ is not Fredholm}\} \quad \text{(essential spectrum of } A\text{)}.$$

Note that $\sigma_e(A) \subset \sigma(A)$. For a semi-Fredholm operator A let

$$\operatorname{ind}(A) := \dim \ker A - \dim(\mathcal{H}/\operatorname{Rng} A)$$

be the *index* of A. When the set $\mathbb{Z} \cup \{\pm\infty\}$ is endowed with the discrete topology, the map $A \mapsto \operatorname{ind}(A)$ (from the set of semi-Fredholm operators to $\mathbb{Z} \cup \{\pm\infty\}$) is continuous [19, p. 361].

8.2 Essential spectrum

We now compute the essential spectrum of

$$T := S|\mathcal{M},$$

where S is, as always, $Sf = zf$ on $H^2(G)$, and $\mathcal{M}$ is a non-zero invariant subspace for S. For $\lambda \in G$, it is easy to show that

$$(S - \lambda I)H^2(G) = \{f \in H^2(G) : f(\lambda) = 0\}$$

and thus is a (closed) non-trivial subspace of $H^2(G)$. Furthermore, $\ker(S - \lambda I) = \{0\}$. From here it follows that $\sigma(S) = \mathbb{D}^-$ and $\sigma_e(S) \subset \partial G$. A result from [20, Thm. 4.3] proves the other inequality and so

$$\sigma_e(S) = \partial G.$$

For each $\lambda \in G$ there is a $c_\lambda > 0$ so that

$$\|(z - \lambda)f\| \geqslant c_\lambda \|f\| \quad \forall f \in H^2(G)$$

and so this same inequality holds for all $f \in \mathcal{M}$. This inequality says that $T - \lambda I$ has closed range. Clearly $\ker(T - \lambda I) = \{0\}$.

For $\lambda \in G \setminus Z(\mathcal{M})$, we can use the nearly invariance of $\mathcal{M}$ (Corollary 6.1.3) to get that

$$(T - \lambda I)\mathcal{M} = \{f \in \mathcal{M} : f(\lambda) = 0\},$$

which is closed. Furthermore, $\mathcal{M}/(T - \lambda I)\mathcal{M}$ is one-dimensional[1]. From here it follows that

$$\sigma(T) = \mathbb{D}^-.$$

By our discussion above, $T - \lambda I$ is Fredholm for all $\lambda \in G \setminus Z(\mathcal{M})$ and

$$\mathrm{ind}(T - \lambda I) := \dim \ker(T - \lambda I) - \dim(\mathcal{M}/(T - \lambda I)\mathcal{M}) = -1 \quad \forall \lambda \in G \setminus Z(\mathcal{M}).$$

From here[2], one can use the fact that $Z(\mathcal{M})$ is a discrete set to show that $(T - \lambda I)$ is Fredholm for *all* $\lambda \in G$ and so

$$\sigma_e(T) \subset \partial G. \tag{8.2.1}$$

In addition,

$$\mathrm{ind}(T - \lambda I) = -1 \quad \forall \lambda \in G. \tag{8.2.2}$$

Using standard Fredholm theory[3] and the above identity on the index, we have

$$\sigma_e(T) = \sigma_l(T), \tag{8.2.3}$$

where $\sigma_l(T)$ is the *left spectrum* (also known as the *approximate point spectrum*) of T. It is a standard fact [19, p. 215] that $\partial \sigma(T) \subset \sigma_l(T)$ and so, since $\sigma(T) = \mathbb{D}^-$, we get

$$\mathbb{T} \subset \sigma_e(T).$$

---

[1] See also (see [59, Lemma 2.1])

[2] We are using the following general fact [21, p. 357]: Suppose $A \in \mathcal{B}(\mathcal{H})$ is Fredholm. Then there is an $\varepsilon > 0$ such that if $Y \in \mathcal{B}(\mathcal{H})$ with $\|Y\|$ (the operator norm of Y) less than ε, then $A + Y$ is also Fredholm and $\mathrm{ind}(A + Y) = \mathrm{ind}(A)$.

[3] Combine Proposition 4.3, Proposition 4.4, and Proposition 6.10 from [21].

Combine this with (8.2.1) and (8.2.3) to obtain

$$\mathbb{T} \subset \sigma_e(T) = \sigma_l(T) \subset \partial G. \tag{8.2.4}$$

Thus to determine $\sigma_e(T)$, it remains to determine which points in $[0, 1)$ belong to $\sigma_e(T)$.

Theorem 8.2.5. *Let $\mathcal{M}$ be a non-zero invariant subspace of $H^2(G)$ and let $A(\mathcal{M})$ be the set of points $x \in [0, 1)$ with the property that there exists an $f_x \in \mathcal{M}$ such that f/f_x extends to be analytic in a neighborhood of x whenever $f \in \mathcal{M}$. Then with $T := S|\mathcal{M}$ we have*

$$\sigma_e(T) = \partial G \setminus A(\mathcal{M}).$$

Proof. Let $x \in A(\mathcal{M})$. For $y \in G \setminus (Z(\mathcal{M}) \cup f_x^{-1}(\{0\}))$ and close to x we know, since $\mathcal{M}$ is nearly invariant (Corollary 6.1.3), that

$$\frac{f - \frac{f}{f_x}(y)f_x}{z - y} \in \mathcal{M} \quad \forall f \in \mathcal{M}.$$

Since $x \in A(\mathcal{M})$ we can let $y \to x$ to obtain

$$Rf := \frac{f - \frac{f}{f_x}(x)f_x}{z - x} \in \mathcal{M} \quad \forall f \in \mathcal{M}.$$

To show that R is continuous on $\mathcal{M}$ we will use the closed graph theorem. Indeed suppose $(f_n)_{n \geqslant 1} \subset \mathcal{M}$ with $f_n \to f$ and $Rf_n \to g$ in the norm of $H^2(G)$. Note that $g \in \mathcal{M}$ and $f_n \to f, Rf_n \to g$ pointwise in G. A little algebra shows that

$$\frac{f_n}{f_x}(x) \to \frac{f}{f_x}(z) - (z - x)\frac{g}{f_x}(z) \quad \forall z \in G \setminus f_x^{-1}(\{0\}).$$

Since $f, g \in \mathcal{M}$, the function on the right is analytic near x and equal to $(f/f_x)(x)$ when $z = x$. Thus

$$\frac{f_n}{f_x}(x) \to \frac{f}{f_x}(x)$$

which says that $g = Rf$ and so, by the closed graph theorem, R is continuous. A routine computation will show that $R(T - xI) = I$. Thus $(T - xI)$ is left-invertible. But since T satisfies $\sigma_l(T) = \sigma_e(T)$ (see (8.2.3)) we see that $x \notin \sigma_e(T)$.

Conversely, suppose that $x \in [0, 1) \setminus \sigma_e(T)$. Then, from (8.2.3), $x \notin \sigma_l(T)$ and so $T - xI$ has a left inverse R (which we will show in a moment is equal to the operator R from the previous paragraph). Since $R(T - xI) = I$ we see that $\text{Rng } R = \mathcal{M}$. Using the fact that $x \notin \sigma_e(T)$ and $\sigma_e(T) \subset \partial G$ (see (8.2.1)), we know, from (8.2.2) and the continuity of the index [19, p. 361], that

$$\text{ind}(T - xI) = \lim_{\varepsilon \to 0^+} \text{ind}(T - (x + i\varepsilon)I) = -1.$$

Furthermore [19, p. 363],

$$
\begin{aligned}
0 &= \mathrm{ind}(R(T - xI)) \\
&= \mathrm{ind}(R) + \mathrm{ind}(T - xI) \\
&= \dim(\ker R) - \dim(\mathrm{Rng}\,R)^{\perp} - 1 \\
&= \dim(\ker R) - 1 \quad (\text{since } \mathrm{Rng}\,R = \mathcal{M}).
\end{aligned}
$$

Thus

$$
\ker R = \mathbb{C} f_x \tag{8.2.6}
$$

for some $f_x \in \mathcal{M} \setminus \{0\}$. For y in some open neighborhood of x, consider the operator

$$
R_y := R(I - (y - x)R)^{-1}.
$$

A computation using the identity

$$
R_y = \sum_{n=0}^{\infty} (y - x)^n R^{n+1} \tag{8.2.7}
$$

(which is valid for y in some small open neighborhood of x) shows that

$$
R_y(T - yI) = R_y((T - xI) + (x - y)I) = I.
$$

Moreover, for $y \in G \setminus (Z(\mathcal{M}) \cup f_x^{-1}(\{0\}))$ and near x, the operator

$$
Q_y f := \frac{f - \frac{f}{f_x}(y) f_x}{z - y} \tag{8.2.8}
$$

is a bounded operator on $\mathcal{M}$ (again using the nearly invariance of $\mathcal{M}$ and the closed graph theorem). A computation will show this operator is also a left inverse for $T - yI$. Since $y \in G \setminus (Z(\mathcal{M}) \cup f_x^{-1}(\{0\}))$, we can use the nearly invariance of $\mathcal{M}$, along with the facts

$$
(T - yI)\mathcal{M} = \{ f \in \mathcal{M} : f(y) = 0 \};
$$

$$
\dim(\mathcal{M}/(T - yI)\mathcal{M}) = 1;
$$

$$
f_x(y) \neq 0,
$$

to see that

$$
\mathcal{M} = (T - yI)\mathcal{M} + \mathbb{C} f_x.
$$

Since $Q_y(T - yI) = R_y(T - yI) = I$, then $Q_y = R_y$ on $(T - yI)\mathcal{M}$. Furthermore, $Q_y f_x = R_y f_x = 0$ (This follows from (8.2.8), (8.2.7), and (8.2.6)). Thus $Q_y = R_y$ on all of $\mathcal{M}$ and so

$$
R_y f = \frac{f - \frac{f}{f_x}(y) f_x}{z - y} \quad \forall f \in \mathcal{M}. \tag{8.2.9}
$$

Finally, the identity in (8.2.7) shows that the function

$$y \mapsto R_y$$

is an operator-valued analytic function for all y in some open neighborhood of x and so for fixed $z_0 \in G$ and $f \in \mathcal{M}$

$$y \to R_y f(z_0)$$

is analytic in some open neighborhood of x. Choosing z_0 such that $f_x(z_0) \neq 0$ and using the identity in (8.2.9), we see that f/f_x is analytic in some open neighborhood of x. Thus $x \in A(\mathcal{M})$ which completes the proof. □

Remark 8.2.10. Analytic continuation across boundary points seems to be a reoccurring theme when studying the essential spectra of multiplication (and Toeplitz) operators on certain Banach spaces of analytic functions [9, 11, 20].

Corollary 8.2.11. *If $f \in H^2(G) \setminus \{0\}$ and $T := S|[f]$, then the following hold.*

1. *The function h/f extends to be analytic on $\mathbb{D}$ for every $h \in \mathcal{M}$.*

2. *$\sigma_e(T) = \mathbb{T}$.*

Proof. Notice how statement (2) follows immediately from statement (1) since (1) shows that $A([f]) = [0, 1)$. To prove (1) fix an $h \in [f]$. By the definition of $[f]$, there is a sequence of analytic polynomials $(p_n)_{n \geqslant 1}$ such that $p_n f \to h$ in the norm of $H^2(G)$. To show that h/f has an analytic continuation across $[0, 1)$, and thus complete the proof of (1), we will show that the sequence $(p_n)_{n \geqslant 1}$ forms a normal family on $\mathbb{D}$. Notice from Proposition 2.4.13 how $(p_n)_{n \geqslant 1}$ forms a normal family on G.

To this end, fix $r \in (0, 1)$ and let $C_r := \{|z| = r\}$. We will assume that r is chosen so that f is non-zero on C_r and that both $f^+(r)$ and $f^-(r)$ exist and are non-zero. It follows that

$$0 < m \leqslant |f(z)| \leqslant M < \infty \quad \forall z \in C_r \cap G.$$

For a compact set $A \subset r\mathbb{D}$ we can apply the Cauchy integral formula (with an appropriate branch cut for the square root) to get

$$(z - r)^{1/2} p_n(z) = \frac{1}{2\pi i} \oint_{C_r} \frac{(\eta - r)^{1/2} p_n(\eta)}{\eta - z} d\eta \quad \forall z \in A.$$

Now apply the following three inequalities

$$|p_n(\eta)| \leqslant \frac{1}{m} |p_n(\eta) f(\eta)|, \quad \eta \in C_r \cap G;$$

$$|p_n(\eta) f(\eta)| \leqslant K \frac{\|p_n f\|}{\operatorname{dist}(\eta, \partial G)^{1/2}}, \quad \eta \in C_r \cap G;[4]$$

---

[4]This inequality follows from Proposition 2.4.13 and (6.2.5).

$$\frac{|\eta - r|^{1/2}}{\operatorname{dist}(\eta, \partial G)^{1/2}} \leqslant K, \quad \eta \in C_r \cap G$$

to the above integral identity to show that

$$|z - r|^{1/2}|p_n(z)| \leqslant K\|p_n f\| \quad \forall z \in A.$$

But since $\|p_n f\|$ is uniformly bounded in n and since $z \in A$ and A is a compact subset of $r\mathbb{D}$, we see that

$$|p_n(z)| \leqslant K \quad \forall z \in A, \quad \forall n \geqslant 1.$$

It follows that the sequence $(p_n)_{n \geqslant 1}$ forms a normal family on $\mathbb{D}$. $\qquad\square$

Corollary 8.2.12. *Let $\mathcal{M}$ be an invariant subspace of $H^2(G)$ and $T = S|\mathcal{M}$. If $x \in [0,1)$ is a cluster point for the zeros or poles of f/g for some $f,g \in \mathcal{M} \setminus \{0\}$, then $x \in \sigma_e(T)$.*

Proof. Assume to the contrary that $x \notin \sigma_e(T)$. Then, by Theorem 8.2.5, there is a function $f_x \in \mathcal{M}$ such that h/f_x extends to be analytic in a neighborhood of x for every $h \in \mathcal{M}$. In particular, f/f_x and g/f_x extend to be analytic near x. Depending on the orders of the zeros of these two functions at x, either f/g or g/f extend to be analytic near x. This contradicts the fact that x is an accumulation point for the zeros (or poles) of f/g. $\qquad\square$

From Theorem 7.2.2 we know, for certain f, that $[f] = \mathcal{M}(\rho, [0,1])$, where $\rho = f^+/f^-$. From here, one might be tempted to conclude that spaces of the form $\mathcal{M}(\rho, [0,1])$, for some measurable $\rho : [0,1] \to \mathbb{C}$, are always cyclic. The following example shows that this is not always the case.

Example 8.2.13. There are two G-inner functions f,g and a measurable function $\rho : [0,1] \to \mathbb{C}$ such that if $\mathcal{M} := [f,g]$, the invariant subspace generated by f and g, then the following hold:

1. $\mathcal{M} \subseteq \mathcal{M}(\rho, [0,1])$.

2. $\sigma_e(S|\mathcal{M}) = \sigma_e(S|\mathcal{M}(\rho, [0,1])) = \partial G$.

3. $S|\mathcal{M}$ and $S|\mathcal{M}(\rho, [0,1])$ are not cyclic.

Proof. Choose a sequence $(a_n)_{n \geqslant 1} \subseteq \mathbb{D}_+$ such that $(a_n)_{n \geqslant 1}$ clusters precisely on all of $[0,1]$ and $(a_n)_{n \geqslant 1}$ is an $H^2(G)$ zero set. Let f be a G-Blaschke product whose zero set is precisely $(a_n)_{n \geqslant 1}$. Since $(a_n)_{n \geqslant 1}$ is also an $H^2(\mathbb{D}_+)$ zero set, then there is also a $\mathbb{D}_+$-Blaschke product b whose zeros are precisely $(a_n)_{n \geqslant 1}$. Notice that b extends to be analytic across $(-1,0)$ since the zeros of b do not accumulate $(-1,0)$. Furthermore, since $|b(x)|^2 = 1$ almost everywhere on $[-1,1]$, the analytic continuation of b across $(-1,0)$ is the function $1/b^*$ where b^* is the $\mathbb{D}_-$-inner function $b^*(z) = \overline{b(\bar{z})}$. Define a function g as follows:

$$g(z) = \begin{cases} f(z)/b(z), & \text{if } z \in \mathbb{D}_+; \\ f(z)b^*(z), & \text{if } z \in \mathbb{D}_-. \end{cases}$$

Notice that $g \in H^\infty(G)$ and that g is a G-inner function. Since b is $\mathbb{D}_+$-inner we also have

$$\frac{g^+(x)}{g^-(x)} = \frac{f^+(x)/b(x)}{f^-(x)b^*(x)} = \frac{f^+(x)b^*(x)}{f^-(x)b^*(x)} = \frac{f^+(x)}{f^-(x)}$$

for almost every $x \in (0,1)$. Thus if we set

$$\rho = g^+/g^- = f^+/f^-,$$

then $\rho : [0,1] \to \mathbb{C}$ is a measurable function and $f,g \in \mathcal{M}(\rho,[0,1])$, so

$$\mathcal{M} := [f,g] \subseteq \mathcal{M}(\rho,[0,1]).$$

Since f/g has infinitely many zeros and poles clustering on $[0,1]$, we have the containment $[0,1] \subseteq \sigma_e(S|\mathcal{M})$ (Corollary 8.2.12). Similarly, since $f,g \in \mathcal{M}(\rho,[0,1])$, we see that $[0,1] \subseteq \sigma_e(S|\mathcal{M}(\rho,[0,1]))$. Thus, from (8.2.4), we have $\sigma_e(S|\mathcal{M}) = \partial G$ and

$$\sigma_e(S|\mathcal{M}(\rho,[0,1])) = \partial G.$$

Hence, by Corollary 8.2.11, neither $S|\mathcal{M}$ nor $S|\mathcal{M}(\rho,[0,1])$ is cyclic. $\qquad\square$

Chapter 9

Other applications

9.1 Compressions

We now examine the compression of S to certain co-invariant subspaces. Throughout this section ω will denote the harmonic measure for ∂G at some point in G. Note from (6.2.5) that $d\omega \asymp |\xi|^{-1/2}|\xi - 1|ds$. We begin with the following.

Proposition 9.1.1. *The map $R : H^2(G) \to L^2([0,1], \omega)$ defined by $Rf = f^+ - f^-$ is a continuous onto linear operator.*

Proof. The obvious estimates will show that R is continuous. Let $\phi = \phi_G : \mathbb{D} \to G$ from the appendix and $\psi = \phi^{-1}$. To show R is onto, let $g \in L^2([0,1], \omega)$ and note that

$$g(x) = k \circ \psi(x)$$

for some $k \in L^2(J, d\theta)$, where $J = \{e^{i\theta} : 0 \leqslant \theta \leqslant \pi/2\}$ and $\psi^+([0,1]) = J$. In other words, we are thinking of g as living on the "top part" of the slit $[0,1]$.

The function $k(e^{i\theta})$, extended to be zero for $\theta \in [\pi/2, \pi]$, has a Fourier sine series

$$k \sim \sum_{n=1}^{\infty} a_n \sin(n\theta),$$

where

$$a_n = \frac{2}{\pi} \int_0^{\pi/2} k(e^{it}) \sin(nt)dt.$$

Define

$$h(z) := \frac{1}{2i} \sum_{n=1}^{\infty} a_n z^n$$

and notice that $h \in H^2(\mathbb{D})$ (since the a_n's are square summable – see (2.1.8)) and

$$h(e^{i\theta}) - h(e^{-i\theta}) = k(e^{i\theta}), \quad \text{a.e. } \theta \in [0, \pi/2].$$

Now let $f := h \circ \psi$ and see that $f \in H^2(G)$ (since $f \circ \phi = h \in H^2(\mathbb{D})$) and for almost every $x \in [0, 1]$,

$$f^+(x) - f^-(x) = h(e^{i\theta}) - h(e^{-i\theta}) = k(e^{i\theta}) = (k \circ \psi)(x) = g(x).$$

Thus R is onto. $\square$

Notice that, $\ker(R) = \mathcal{M}(1)$. This allows us to define the quotient operator

$$\widetilde{R} : H^2(G)/\mathcal{M}(1) \to L^2([0, 1], \omega), \quad \widetilde{R}\widetilde{f} := Rf = f^+ - f^-,$$

where $\widetilde{f}$ is the coset in $H^2(G)/\mathcal{M}(1)$ represented by f. With S defined as multiplication by z on $H^2(G)$, one forms the bounded operator

$$\widetilde{S} : H^2(G)/\mathcal{M}(1) \to H^2(G)/\mathcal{M}(1), \quad \widetilde{S}\widetilde{f} := \widetilde{zf}.$$

An easy calculation shows that

$$\widetilde{R}\widetilde{S} = M_x \widetilde{R}, \tag{9.1.2}$$

where $M_x g = xg$ is multiplication by x on $L^2([0, 1], \omega)$. Putting the above discussion in more appropriate Hilbert space language yields the following.

Proposition 9.1.3. *Suppose $\mathcal{N}$ is the co-invariant subspace $\mathcal{N} = H^2(G) \ominus \mathcal{M}(1)$, $P_{\mathcal{N}}$ is the orthogonal projection of $H^2(G)$ onto $\mathcal{N}$, and $C = P_{\mathcal{N}} S|\mathcal{N}$ is the compression of S to $\mathcal{N}$. Then C is similar to M_x on $L^2([0, 1], \omega)$.*

Corollary 9.1.4. *If $h, 1/h \in H^\infty(G)$ and $\rho = h^+/h^-$, then the compression of S to $H^2(G) \ominus M(\rho)$ is similar to M_x on $L^2([0, 1], \omega)$.*

Proof. Let $A = RM_{1/h}$ where R is the map from Proposition 9.1.1 and the operator $M_{1/h} : H^2(G) \to H^2(G)$ is defined by $M_{1/h}f = f/h$. One easily sees that $A : H^2(G) \to L^2([0, 1], \omega)$ is onto, intertwines S with M_x, and the kernel of A is precisely $M(\rho)$. The result follows by passing to quotients as above. $\square$

Using Wiener's theorem [43, p. 7] we know that every M_x-invariant subspace of $L^2([0, 1], \omega)$ is of the form $\chi_{E^c}L^2([0, 1], \omega)$ for some measurable set $E \subset [0, 1]$. Bringing in Theorem 7.2.2 and (9.1.2), we can prove the following.

Theorem 9.1.5. *Suppose $\mathcal{M}$ is an invariant subspace of $H^2(G)$ and there exists an $f \in \mathcal{M}$ such that f and $1/f$ belong to $H^\infty(G)$. Then there is a measurable set $E \subset [0, 1]$ such that $\mathcal{M} = \mathcal{M}(\rho, E)$, where $\rho = f^+/f^-$.*

Remark 9.1.6. Compare this to Theorem 7.2.2.

9.2 The parameters

Let us focus a bit more on the invariant subspaces

$$\mathcal{M}(\rho) := \left\{ f \in H^2(G) : f^+ = \rho f^- \text{ a.e. on } [0,1] \right\},$$

where ρ is a complex-valued measurable function on $[0,1]$. We ask the question: For what measurable ρ is $\mathcal{M}(\rho) \neq \{0\}$?

Let us first discuss a necessary condition. We see from (2.4.3) that for $F \in H^2(G) \setminus \{0\}$,

$$\int_{[0,1]} \left(|\log|F^+|| + |\log|F^-|| \right) d\omega + \int_{\mathbb{T}} |\log|F|| d\omega < \infty.$$

In particular, if $\rho : [0,1] \to \mathbb{C}$ is measurable and $F \in \mathcal{M}(\rho) \setminus \{0\}$, we can combine the above observation with the identity $\rho = F^+/F^-$ almost everywhere to see the following.

Proposition 9.2.1. *Suppose* $\rho : [0,1] \to \mathbb{C}$ *is measurable and* $\mathcal{M}(\rho) \neq \{0\}$, *then* $\log|\rho| \in L^1([0,1], \omega)$.

Is the necessary condition in the above proposition sufficient? Let us work through a few examples.

Proposition 9.2.2. *If* ρ *is a complex-valued step function which is never zero on* $[0,1]$, *then* $\mathcal{M}(\rho) \neq \{0\}$.

Proof. Let $p_0, p_1, \ldots, p_n$ be non-zero complex numbers such that $\Re p_j \geq 0$ for all j and

$$0 \leq a_1 < a_2 < \cdots < a_n < 1.$$

Let

$$f(z) := z^{p_0}(z - a_1)^{p_1} \cdots (z - a_n)^{p_n},$$

where we take the branch of the complex logarithm to be analytic on $\mathbb{C} \setminus [0, \infty)$. Notice, since $\Re p_j \geq 0$, that $f \in H^\infty(G) \setminus \{0\}$. Then with

$$\rho(x) := e^{2\pi i p_0}, \quad 0 < x < a_1,$$

$$\rho(x) := e^{2\pi i p_0} e^{2\pi i p_1}, \quad a_0 < x < a_1,$$

and so on, we see that

$$f^+ = \rho f^-$$

and so $\mathcal{M}(\rho) \neq \{0\}$. By choosing appropriate a_j and p_j one can create any non-vanishing step function ρ. $\qquad\square$

Proposition 9.2.3. *Suppose* ρ *is a real-valued measurable function on* $[0,1]$ *such that*

$$a \leq \rho(x) \leq b, \quad x \in [0,1] \tag{9.2.4}$$

for some constants $0 < a < b < \infty$. *Then* $\mathcal{M}(\rho) \neq \{0\}$.

Proof. Let

$$h(z) := \frac{1}{2\pi i} \int_0^1 \frac{\log \rho(t)}{t - z} dt$$

be the Borel transform of the function $\chi_{[0,1]} \log \rho$ and notice that h is analytic on $\mathbb{C} \setminus [0,1]$. It is well known that the limits $h^+(x)$ and $h^-(x)$ exist for almost every $x \in \mathbb{R}$ (see [70, Theorem 1.4] for details).

A computation shows that

$$h(x + iy) - h(x - iy) = \int_0^1 P_{x+iy}(t) \log \rho(t) dt,$$

where

$$P_{x+iy}(t) = \frac{1}{\pi} \frac{y}{(t-x)^2 + y^2}$$

is the usual Poisson kernel for the upper half plane. We will allow negative values of y in the above formula. Using standard facts about Poisson integrals [34, p. 29], we get

$$h^+ - h^- = \chi_{[0,1]} \log \rho$$

almost everywhere. The identities

$$P_{x+iy}(t) = -P_{x-iy}(t)$$

and

$$\Re h(x + iy) = \frac{1}{2} \int_0^1 P_{x+iy}(t) \log \rho(t) dt$$

along with (9.2.4) will show that $\Re h$ is a bounded function on G. Thus $f := e^h$ belongs to $H^\infty(G) \setminus \{0\}$ and $f^+ = \rho f^-$ and so $f \in \mathcal{M}(\rho) \setminus \{0\}$. $\qquad \square$

Chapter 10

Domains with several slits

10.1 Statement of the result

We now extend our main theorem (Theorem 6.2.1) to domains with several slits. More precisely, we consider domains of the form

$$G = \mathbb{D} \setminus \bigcup_{j=1}^{N} \gamma_j, \qquad (10.1.1)$$

where γ_j are analytic arcs satisfying certain technical conditions. They are the following (see Figure 10.1 for an example):

1. Each γ_j is a closed connected subset of a simple analytic open arc $\widehat{\gamma}_j$ which meets $\mathbb{T}$ at a positive angle. The arc $\widehat{\gamma}_j$ will be called an *analytic continuation* of γ_j.

2. $\gamma_1, \ldots, \gamma_N$ are pairwise disjoint.

3. Each γ_j has one end point $\lambda_j \in \mathbb{T}$ and $\gamma_j \setminus \{\lambda_j\} \subset \mathbb{D}$.

Our extension of Theorem 6.2.1 is the following.

Theorem 10.1.2. *Let G be a domain as in (10.1.1) and let $\widehat{\gamma}_j, 1 \leqslant j \leqslant N$, be analytic continuations of the arcs $\gamma_j, 1 \leqslant j \leqslant N$. If $\mathcal{M}$ is a non-trivial invariant subspace of $H^2(G)$ with greatest common G-inner divisor $\Theta_{\mathcal{M}}$, then there is a measurable set*

$$E \subset \bigcup_{j=1}^{N} \gamma_j,$$

a measurable function $\rho : E \to \mathbb{C}$, and an analytic function F on an open set $V \subset G$, with

$$\bigcup_{j=1}^{N} \widehat{\gamma}_j \cap \mathbb{D} \subset V,$$

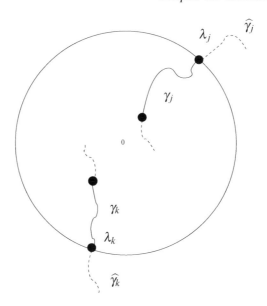

Figure 10.1: A domain as in (10.1.1) with several slits. Notice how each slit γ_j is part of a larger analytic slit $\widehat{\gamma}_j$ and how the slits meet the circle at a positive angle.

such that

$$\mathcal{M} = \Theta_{\mathcal{M}} \cdot \left\{ f \in H^2(G) : \frac{f|V}{F} \in H^2(V), f^+ = \rho f^- \ a.e. \ on \ E \right\}.$$

Remark 10.1.3.   1. The functions f^+ and f^- are the non-tangential limits of f from opposite sides (once an orientation is fixed) of the arc γ_j.

2. The open set V will turn out to be

$$V = \bigcup_{j=1}^{N}(V_j \cap G),$$

where V_j are certain disjoint domains obtained from Lemma 10.2.1 (below). See also Figure 10.2.

3. Since V is a disjoint union of simply connected domains, we should be clear what we mean by $H^2(V)$. We follow [20, p. 664]. An analytic function f on V belongs to $H^2(V)$ if $|f|^2$ has a harmonic majorant on V. We let U_f denote the least harmonic majorant and for $a_j \in V_j \cap G$, $1 \leqslant j \leqslant N$, define the norm on $H^2(V)$ to be

$$\|f\|_{H^2(V)}^2 := \sum_{j=1}^{N} U_f(a_j).$$

As to be expected, different choices of a_j's yield equivalent norms.

4. It should be the case, as in Theorem 6.2.1, that there is a $\widehat{G}$-outer function F, where

$$\widehat{G} = \mathbb{D} \setminus \bigcup_{j=1}^{N} (\widehat{\gamma_j})^{-},$$

such that

$$\mathcal{M} = \Theta_{\mathcal{M}} \cdot \left\{ f \in H^2(G) : \frac{f}{F} \in H^2(\widehat{G}), f^{+} = \rho f^{-} \text{ a.e. on } E \right\}.$$

However, at this point, we do not see how to produce this more global outer function from the local function we have in Theorem 10.1.2.

10.2   Some technical lemmas

We will use Theorem 6.2.1 to prove Theorem 10.1.2. In order to do this, we need to take care of a few technicalities.

Lemma 10.2.1. *There are open subsets $V_1, \ldots, V_N$ of $\mathbb{D}$ such that*

1. *$V_1, \ldots, V_N$ are pairwise disjoint.*

2. *For each j, V_j contains $\gamma_j \setminus \{\lambda_j\}$.*

3. *For each j, $\partial V_j \setminus \{\lambda_j\}$ is a C^2 arc and ∂V_j is a piecewise C^2 arc which intersects $\mathbb{T}$ only at λ_j and, at this point, makes a positive angle with both γ_j and $\mathbb{T}$.*

4. *For each j, there is a conformal map β_j from V_j onto $\mathbb{D}$ such that*

$$\beta_j(\gamma_j \setminus \{\lambda_j\}) = [0,1).$$

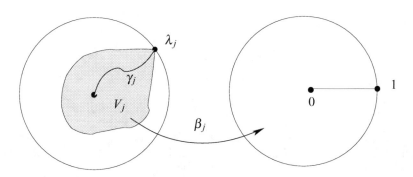

Figure 10.2: The domain V_j (shaded) and the conformal map $\beta_j : V_j \to \mathbb{D}$ which satisfies $\beta_j(\gamma_j \setminus \{\lambda_j\}) = [0,1)$.

Proof of Lemma 10.2.1. Fix $j = 1, \ldots, N$ and let $\gamma = \gamma_j$, $\widehat{\gamma} = \widehat{\gamma}_j$, and $\lambda = \lambda_j$. By the definition of analytic arc, there is an open set U of $\widehat{\gamma}$ and a conformal map α which maps U onto a region R which is symmetric about an open interval (a, b) which contains $[0, 1]$. We can assume that $\alpha(\gamma) = [0, 1]$ and $\alpha(\lambda) = \{1\}$. By shrinking U, we can also assume that ∂R is analytic (see Figure 10.3).

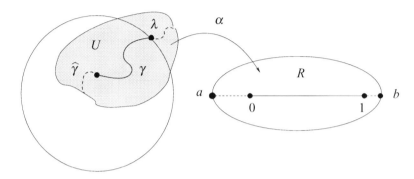

Figure 10.3: The region U (shaded) and $R = \alpha(U)$. Note that $\alpha(\gamma) = [0, 1]$ with $\alpha(\lambda) = \{1\}$.

The curve $\alpha(U \cap \mathbb{T})$ will be an analytic arc in R that passes through the point 1. Pick two points $z_1, z_2 \in \partial R$ which are symmetric about $[a, b]$ and such that the line segments ℓ_1, ℓ_2 (which connect z_1, respectively z_2, to 1) only intersect $Y := \alpha(U \cap \mathbb{T})$ at 1 and form a positive acute angle to both $[a, b]$ and Y at 1. Now form the region R_1 bounded by the line segment $[a, 1]$, the line ℓ_1, and the part of ∂R subtended by z_1 and a (see Figure 10.4).

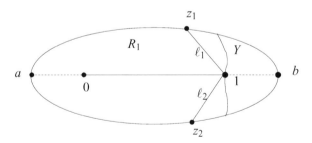

Figure 10.4: The region R_1 (top).

By altering the definition of R_1 slightly (replacing part of the line segments ℓ_1 and ℓ_2 with appropriate C^2 curves near the points z_1 and z_2) we can assume that $\partial R_1 \setminus \{1\}$ is C^2.

One can find a conformal map α_1 from R_1 onto $\mathbb{D}_+$ such that

$$\alpha_1(a) = -1, \quad \alpha_1(0) = 0, \quad \alpha_1(1) = 1$$

(see [55, p. 319]). By the symmetry principle, α_1 (really the analytic continuation of α_1) will map $R_1 \cup (a,1) \cup \{\bar{z} : z \in R_1\}$ onto $\mathbb{D}$ with $\alpha_1([0,1)) = [0,1)$. Therefore the map $\alpha^{-1} \circ \alpha_1^{-1}$ will map $\mathbb{D}$ onto a region $V = V_j$ with the desired properties (2) and (3). Finally, let $\beta = \beta_j$ be the inverse of $\alpha^{-1} \circ \alpha_1^{-1}$. $\qquad\square$

10.3 A localization of Yakubovich

The proof of Theorem 10.1.2 depends on this following localization result discovered by Yakubovich [78, Lemma 4] in the case of an annular domain.

Proposition 10.3.1. *Suppose that G is a multiple slit domain as in* (10.1.1) *and let*

$$V := \bigcup_{j=1}^{N} (V_j \cap G).$$

If $\mathcal{M}$ is a non-trivial invariant subspace of $H^2(G)$ with greatest common G-inner divisor equal to one, then $f \in H^2(G)$ belongs to $\mathcal{M}$ if and only if $f|V$ belongs to the closure of $\mathcal{M}|V$ in $H^2(V)$.

To make writing integrals more manageable, we will let ω be the harmonic measure for G at $\phi(0)$, where $\phi : \mathbb{D} \to G$ and $\psi = \phi^{-1}$. From our discussion of the harmonic measure from Chapter 2, note that

$$\omega = \sum_{j=1}^{N} (\omega_{0,\mathbb{D}} \circ \psi_j^+ + \omega_{0,\mathbb{D}} \circ \psi_j^-) + \omega_{0,\mathbb{D}} \circ \psi|\mathbb{T}$$

$$= \sum_{j=1}^{N} \left(\frac{|(\psi')_j^+|}{2\pi} ds + \frac{|(\psi')_j^-|}{2\pi} ds \right) + |\psi'| \frac{d\theta}{2\pi}.$$

In the above, $(\psi')_j^+$ and $(\psi')_j^-$ denote the upper and lower boundary functions for ψ' on γ_j. For $h \in H^2(G) \cdot \overline{H^2(G)}$, we will use the notation

$$\int_{\partial G} h d\omega^* := \sum_{j=1}^{N} \left(\int_{\gamma_j} h_j^+ \frac{|(\psi')_j^+|}{2\pi} ds + \int_{\gamma_j} h_j^- \frac{|(\psi')_j^-|}{2\pi} ds \right) + \int_{\mathbb{T}} h|\psi'| \frac{d\theta}{2\pi}. \qquad (10.3.2)$$

We invite the reader to verify the following Cauchy-Schwarz type inequality

$$\int_{\partial G} |f\bar{g}| d\omega^* \leqslant (2N+1) \left(\int_{\partial G} |f|^2 d\omega^* \right)^{1/2} \left(\int_{\partial G} |g|^2 d\omega^* \right)^{1/2}. \qquad (10.3.3)$$

Needed in the proof of Proposition 10.3.1 will be the following discussion of the growth (and decay) of inner and outer functions. If S is a singular $\mathbb{D}$-inner function with an atom at $\zeta_0 \in \mathbb{T}$, then

$$|S(r\zeta_0)| = O(e^{-\frac{c}{1-r}}), \quad r \to 1^-$$

for some $c > 0$. For $g \in L^1(m)$ we have

$$\int_{\mathbb{T}} P_{r\zeta_0}(\xi)|g(\xi)|dm(\xi) = o(\frac{1}{1-r}), \quad r \to 1^-. \tag{10.3.4}$$

Certainly (10.3.4) is true when g is continuous. To see the general case, approximate g with continuous functions in the $L^1(m)$ norm and use the identity

$$\sup_{\xi \in \mathbb{T}} P_{r\zeta_0}(\xi) = \frac{1+r}{1-r}.$$

This means that if H is a $\mathbb{D}$-outer function, then

$$H(r\zeta_0) = \exp\left(\int_{\mathbb{T}} \frac{\xi + r\zeta_0}{\xi - r\zeta_0} \log|H(\xi)|dm(\xi)\right)$$

and so, since

$$\Re\frac{\xi + r\zeta_0}{\xi - r\zeta_0} = P_{r\zeta_0}(\xi)$$

and $\log|H| \in L^1(m)$, we have, via (10.3.4),

$$\frac{1}{|H(r\zeta_0)|} \leqslant e^{\frac{c_r}{1-r}},$$

where $c_r \to 0$ as $r \to 1^-$. From here we see that

$$\lim_{r \to 1^-}\left|\frac{S(r\zeta_0)}{H(r\zeta_0)}\right| = 0. \tag{10.3.5}$$

See [40] for a related result.

We are now ready for the proof of Proposition 10.3.1.

Proof of Proposition 10.3.1. Recall that the regions V_j are from Lemma 10.2.1 and the open set V is

$$\bigcup_{j=1}^{N}(V_j \cap G).$$

It is obvious that if $f \in \mathcal{M}$, then $f|V \in \mathcal{M}|V$. So suppose that $g \in H^2(G)$ and

$$g|V = \lim_{n \to \infty} f_n|V, \tag{10.3.6}$$

where $f_n \in \mathcal{M}$ and the limit in (10.3.6) is in the norm of $H^2(V)$. Our goal is to show that $g \in \mathcal{M}$.

Recall that λ_j is the endpoint of γ_j which lies on $\mathbb{T}$. Define a polynomial Q by

$$Q(z) = \prod_{j=1}^{N}(z - \lambda_j)^b,$$

where b is a positive integer large enough so that

$$|Q|d\omega^* \leqslant cd\omega_V^* \quad \text{on } \bigcup_j \gamma_j; \tag{10.3.7}$$

and

$$|Q|ds \leqslant cd\omega_V^* \quad \text{on } \bigcup_j \partial V_j. \tag{10.3.8}$$

Here is where we use the hypothesis that $\partial V_j \setminus \{\lambda_j\}$ is C^2 and ∂V_j is a piecewise C^2 arc which meets $\mathbb{T}$ and γ_j at positive angles. See (2.3.14) for an explanation of this.

The rather lengthy argument below will show, for each $h \in \mathcal{M}^\perp$, that

$$\left\langle \frac{Qg}{z-\lambda}, h \right\rangle = 0, \quad |\lambda| > 1. \tag{10.3.9}$$

Assuming this is true, let us see how prove that $g \in \mathcal{M}$. Using (10.3.9) and Runge's theorem [18, p. 198] we see that $Qg \perp h$ for all $h \in \mathcal{M}^\perp$. Hence $Qg \in \mathcal{M}$. We now need to argue that $g \in \mathcal{M}$. Since the zeros of the polynomial Q lie on $\mathbb{T}$, Q is $\mathbb{D}$-outer and so [34, p. 85] we can choose $(Q_n)_{n\geqslant 1} \subset H^\infty(\mathbb{D})$ such that $Q_nQ \to 1$ weak-$*$ in $H^\infty(\mathbb{D})$. Since $\mathcal{M}$ is $H^\infty(\mathbb{D})$-invariant (Proposition 6.1.1), we see that $Q_nQg \in \mathcal{M}$. Furthermore, $Q_nQg \to g$ weakly in $H^2(G)$ and so $g \in \mathcal{M}$.

Before getting to the heart of the proof of (10.3.9), we first need to derive a few integral formulas. For $f \in \mathcal{M} \setminus \{0\}$ and $h \in \mathcal{M}^\perp$, note, since $\mathcal{M}$ is invariant, that

$$\left\langle \frac{f}{z-\lambda}, h \right\rangle = 0, \quad |\lambda| > 1. \tag{10.3.10}$$

If

$$w(\zeta) := |\psi'(\zeta)|, \quad \zeta \in \mathbb{T},$$

an application of Fatou's jump theorem (Theorem 3.3.3) implies

$$\lim_{r \to 1^-} \int_{\partial G} \frac{f\bar{h}}{z-r\zeta} d\omega^* = \bar{\zeta} f(\zeta)\overline{h(\zeta)}w(\zeta), \quad \text{a.e. } \zeta \in \mathbb{T}. \tag{10.3.11}$$

Recall our notational understanding discussed just before this proof (see (10.3.2)) and notice in (10.3.11) how the integrals over the slits cancel out in the limit. Define

$$h_1(\lambda) := \frac{1}{f(\lambda)} \int_{\partial G} \frac{f\bar{h}}{z-\lambda} d\omega^*, \quad \lambda \in G \setminus f^{-1}(\{0\}). \tag{10.3.12}$$

From (10.3.11) we get

$$\lim_{r \to 1^-} h_1(r\zeta) = \bar{\zeta}\overline{h(\zeta)}w(\zeta), \quad \text{a.e. } \zeta \in \mathbb{T}. \tag{10.3.13}$$

This shows, via uniqueness of radial limits of quotients of $H^p(G)$ functions, that h_1 is independent of $f \in \mathcal{M} \setminus \{0\}$. Thus, since the greatest common G-inner divisor of $\mathcal{M}$ is 1, we conclude that h_1, initially defined on $G \setminus f^{-1}(\{0\})$, has an analytic continuation to G and furthermore,

$$h_1 \in N^+(G).$$

Note that $Qf \in \mathcal{M}$ and so from (10.3.12) we have

$$Q(\lambda) f(\lambda) h_1(\lambda) = \int_{\partial G} \frac{Qf\overline{h}}{z - \lambda} d\omega^*. \tag{10.3.14}$$

Since the curves $\partial V_j, \gamma_j, \mathbb{T}$ all meet at λ_j at positive angles to each other, we see that for fixed $\lambda \in \partial V_j \setminus \{\lambda_j\}$

$$\sup_{\zeta \in \mathbb{T}} \frac{1}{|\zeta - \lambda|} \asymp \frac{1}{|\lambda - \lambda_j|},$$

$$\sup_{z \in \gamma_j} \frac{1}{|z - \lambda|} \asymp \frac{1}{|\lambda - \lambda_j|}.$$

From these estimates, (10.3.14), and the Cauchy-Schwarz type inequality from (10.3.3) it follows that Qfh_1 is a bounded function on each ∂V_j and

$$\sup \left\{ |Q(\lambda) f(\lambda) h_1(\lambda)| : \lambda \in \bigcup_{j=1}^N \partial V_j \right\} \leqslant c \|f\|_{H^2(G)}, \quad f \in \mathcal{M}, \tag{10.3.15}$$

where $c > 0$ is independent of f.

For general $g \in H^2(G)$ (not necessarily in $\mathcal{M}$), define

$$\ell(\lambda, g) := \int_{\partial G} \frac{Qg\overline{h}}{z - \lambda} d\omega^*, \quad \lambda \in \mathbb{D}_e,$$

$$\ell(\lambda, g) := \int_{\partial G} \frac{Qg\overline{h}}{z - \lambda} d\omega^* - Q(\lambda) g(\lambda) h_1(\lambda), \quad \lambda \in G.$$

An argument using Fatou's jump theorem (Theorem 3.3.3) and (10.3.13) will show that

$$\lim_{r \to 1^-} \ell(r\zeta, g) = \lim_{s \to 1^+} \ell(s\zeta, g), \quad \text{a.e. } \zeta \in \mathbb{T} \tag{10.3.16}$$

and so the functions $\ell(\cdot, g)|G$ and $\ell(\cdot, g)|\mathbb{D}_e$ are 'pseudocontinuations' of each other across $\mathbb{T}$ (see [63] for more on pseudocontinuations). Privalov's uniqueness theorem says that pseudocontinuations are unique in that if L is another analytic function on G whose non-tangential boundary values are equal almost everywhere to those of $\ell(\cdot, g)|\mathbb{D}_e$, then $L = \ell(\cdot, g)|G$. If $g \in \mathcal{M}$, then by (10.3.10) $\ell(\cdot, g) \equiv 0$ on $\mathbb{D}_e$ and so, by uniqueness of pseudocontinuations,

$$\ell(\cdot, g) \equiv 0 \text{ on } G \cup \mathbb{D}_e \text{ for all } g \in \mathcal{M}. \tag{10.3.17}$$

For a compactly supported measure μ in $\mathbb{C}$, the Cauchy transform

$$(C\mu)(\lambda) := \int \frac{1}{z - \lambda} d\mu(z)$$

is analytic on $\widehat{\mathbb{C}} \setminus \mathrm{supt}(\mu)$ and

$$|(C\mu)(\lambda)| \leqslant \frac{1}{\mathrm{dist}(\lambda,\mathrm{supt}(\mu))}\|\mu\|, \quad \lambda \in \widehat{\mathbb{C}} \setminus \mathrm{supt}(\mu), \tag{10.3.18}$$

where $\|\mu\|$ is the total variation norm of μ. For a compact set $K \subset G$, we can use the estimate in (10.3.18), the standard continuity of point evaluations for $H^2(G)$ (Proposition 2.4.13), and (10.3.3) to obtain a constant $C_K > 0$, depending only on K, with

$$|\ell(\lambda,g)| \leqslant C_K \|g\|_{H^2(G)} \quad \forall g \in H^2(G), \lambda \in K. \tag{10.3.19}$$

We will now derive, for certain $g \in H^2(G)$, a useful integral formula for $\ell(\cdot,g)$. Define

$$P_V := \{g \in H^2(G) : Qgh_1|(\cup_j \partial V_j) \in L^1(\cup_j \partial V_j, ds)\}$$

and notice from (10.3.15) that

$$\mathcal{M} \subset P_V. \tag{10.3.20}$$

Define

$$\Omega := G \setminus V^-.$$

For $\lambda \in V$ and $g \in P_V$, the function

$$H_\lambda(z) := \frac{Q(z)g(z)h_1(z)}{z - \lambda}$$

belongs to $N^+(\Omega)$. Furthermore, by (10.3.13),

$$H_\lambda(\zeta) = \frac{Q(\zeta)g(\zeta)\overline{\zeta h(\zeta)}w(\zeta)}{\zeta - \lambda} \quad \text{a.e. } \zeta \in \mathbb{T}.$$

Observing that $g|\mathbb{T}, h|\mathbb{T} \in L^2(wdm)$ we can use the Cauchy-Schwarz inequality to see that $H_\lambda|\mathbb{T} \in L^1(m)$. Using this observation along with our assumption that $g \in P_V$ we get that $H_\lambda|\partial\Omega \in L^1(\partial\Omega, ds)$. Now apply Proposition 2.4.10 to see that $H_\lambda \in E^1(\Omega)$. From Proposition 2.4.12, the familiar Cauchy's theorem is now valid and so

$$\oint_{\mathbb{T}} H_\lambda(z)dz - \oint_{\cup_j \partial V_j} H_\lambda(z)dz = \oint_{\partial\Omega} H_\lambda(z)dz = 0, \tag{10.3.21}$$

where the orientation of the path integrals obeys the usual left-hand rule. Thus, for $\lambda \in V$ and $g \in P_V$, we have

$$\ell(\lambda,g) = \int_{\partial G} \frac{Qg\overline{h}}{z-\lambda} d\omega^* - Q(\lambda)g(\lambda)h_1(\lambda)$$

$$= \frac{1}{2\pi i} \oint_{\mathbb{T}} \frac{Qgh_1}{\zeta-\lambda} d\zeta + \int_{\cup_j \gamma_j} \frac{Qg\overline{h}}{z-\lambda} d\omega^* - Q(\lambda)g(\lambda)h_1(\lambda) \quad (d\zeta = i\zeta|d\zeta| \text{ and } (10.3.13))$$

$$= \frac{1}{2\pi i} \oint_{\cup_j \partial V_j} \frac{Qgh_1}{z-\lambda} dz + \int_{\cup_j \gamma_j} \frac{Qg\overline{h}}{z-\lambda} d\omega^* - Q(\lambda)g(\lambda)h_1(\lambda) \quad \text{(by (10.3.21))}$$

For $\lambda \in \mathbb{D}_e$ and $g \in P_V$, a similar computation with Cauchy's theorem yields

$$\ell(\lambda, g) = \frac{1}{2\pi i} \oint_{\cup_j \partial V_j} \frac{Qgh_1}{z - \lambda} dz + \int_{\cup_j \gamma_j} \frac{Qg\overline{h}}{z - \lambda} d\omega^*. \tag{10.3.22}$$

For $\lambda \in G \setminus V^- = \Omega$ we have

$$\int_{\mathbb{T}} \frac{Qg\overline{h}}{\zeta - \lambda} d\omega^*$$

$$= \frac{1}{2\pi i} \oint_{\mathbb{T}} \frac{Qgh_1}{\zeta - \lambda} d\zeta \quad (d\zeta = i\zeta |d\zeta| \text{ and } (10.3.13))$$

$$= \frac{1}{2\pi i} \oint_{\partial \Omega} \frac{Qgh_1}{z - \lambda} dz + \frac{1}{2\pi i} \oint_{\cup_j \partial V_j} \frac{Qgh_1}{z - \lambda} dz$$

$$= Q(\lambda) g(\lambda) h_1(\lambda) + \frac{1}{2\pi i} \oint_{\cup_j \partial V_j} \frac{Qgh_1}{z - \lambda} dz \text{ (Cauchy's formula – Proposition 2.4.12)}$$

which says, for $\lambda \in G \setminus V^-$, that

$$\ell(\lambda, g) = \frac{1}{2\pi i} \oint_{\cup_j \partial V_j} \frac{Qgh_1}{z - \lambda} dz + \int_{\cup_j \gamma_j} \frac{Qg\overline{h}}{z - \lambda} d\omega^*. \tag{10.3.23}$$

Combining (10.3.22) and (10.3.23) we see that $\ell|\mathbb{D}_e$ and $\ell|(G \setminus V^-)$ are analytic continuations of each other. In summary, we have, for $g \in P_V$, that $\ell(\lambda, g)$ (originally defined on $\mathbb{D}_e \cup G$) extends to be an analytic function on $(\widehat{\mathbb{C}} \setminus V^-) \cup V$ which satisfies the formulas

$$\ell(\lambda, g) = \frac{1}{2\pi i} \oint_{\cup_j \partial V_j} \frac{Qgh_1}{z - \lambda} dz + \int_{\cup_j \gamma_j} \frac{Qg\overline{h}}{z - \lambda} d\omega^*, \quad \lambda \in \widehat{\mathbb{C}} \setminus V^-, \tag{10.3.24}$$

while

$$\ell(\lambda, g) = \frac{1}{2\pi i} \oint_{\cup_j \partial V_j} \frac{Qgh_1}{z - \lambda} dz + \int_{\cup_j \gamma_j} \frac{Qg\overline{h}}{z - \lambda} d\omega^* - Q(\lambda) g(\lambda) h_1(\lambda), \quad \lambda \in V. \tag{10.3.25}$$

Suppose Γ is a circle contained in G which intersects $\cup_j \partial V_j$ in a finite set and such that the angle at each point of intersection with $\cup_j \partial V_j$ is different from zero or π. We can apply our Cauchy transform estimate from (10.3.18), along with (10.3.3) and Proposition 2.4.13, to (10.3.24) and (10.3.25) to get

$$\text{dist}(\lambda, \Gamma \cap \partial V) |\ell(\lambda, g)| \leqslant A_\Gamma \left(\|g\|_{H^2(G)} + \int_{\cup_j \partial V_j} |Qgh_1| ds \right), \quad \lambda \in \Gamma, \quad g \in P_V, \tag{10.3.26}$$

where $A_\Gamma > 0$ depends only on Γ.

Let σ be a G-inner function with atomic singularities at $\{\lambda_1, \ldots, \lambda_N\}$ (note that $\{\lambda_j\} = \gamma_j \cap \mathbb{T}$). From the fact that ∂V_j meets $\mathbb{T}$ at a positive angle we see, for every $t > 0$, that $\sigma|\partial V_j$ decreases to zero faster than any G-outer function on ∂V_j (see (10.3.5)). Thus, for all $t > 0$, $\sigma^t h_1$ is a bounded function on ∂V_j for each j. This means that for any $g \in H^2(G)$ and $t > 0$ we have

$$\int_{\cup_j \partial V_j} |Qg\sigma^t h_1| ds \leqslant c_t \int_{\cup_j \partial V_j} |Qg| ds$$

$$\leqslant c_t \int_{\cup_j \partial V_j} |g| d\omega_V^* \quad \text{(by (10.3.8))}$$

$$\leqslant c_t \|g\|_{H^2(V)}$$

$$\leqslant c_t \|g\|_{H^2(G)} \quad \text{(by (2.1.4))}.$$

Hence

$$\sigma^t g \in P_V \quad \forall g \in H^2(G), t > 0. \tag{10.3.27}$$

For each $j = 1, \dots, N$, let Δ_j be an open disk with $\Delta_j^- \subset G$ and such that $\partial \Delta_j \cap \partial V_j$ is a finite set and the angle between $\partial \Delta_j$ and ∂V_j is different from zero or π (see Figure 10.5).

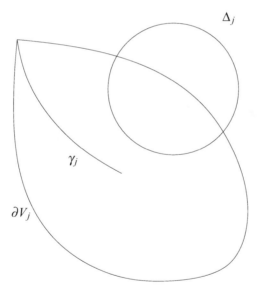

Figure 10.5: The disks Δ_j.

The facts that $\sigma^{1/n} \to 1$ almost everywhere on ∂G and $|\sigma^{1/n} - 1| \leqslant 2$, along with the dominated convergence theorem and (10.3.19), will show that

$$\ell(\lambda, g) = \lim_{n \to \infty} \ell(\lambda, \sigma^{1/n} g) \quad \text{uniformly on } \cup_j \partial \Delta_j. \tag{10.3.28}$$

We pause to comment that (10.3.27) says that the functions $\sigma^{1/n} f$ ($f \in \mathcal{M}$) and $\sigma^{1/n} g$ ($g \in H^2(G)$) can be applied to the integral formulas in (10.3.24) and (10.3.25) as well as the estimate in (10.3.26).

After this long preamble and getting things set up, we are now ready to complete the proof by verifying the identity in (10.3.9). If $f_n \in \mathcal{M}$ approximate g as in (10.3.6), we can use the facts that $|\sigma^{1/n}| \leqslant 1$, $\sigma^{1/n}h_1$ is bounded on $\cup_j \partial V_j$, (10.3.8), and (10.3.3) to assume (by choosing an appropriate subsequence of the f_n's) that

$$\lim_{n \to \infty} \left(\|\sigma^{1/n}(g-f_n)\|_{H^2(V)} + \int_{\cup_j \partial V_j} |Q(g-f_n)\sigma^{1/n}h_1| ds \right) = 0.$$

Apply this last limit identity to (10.3.26) to get

$$\lim_{n \to \infty} \text{dist}(\lambda, \partial \Delta_j \cap \partial V_j) \left| \ell(\lambda, \sigma^{1/n}g) - \ell(\lambda, \sigma^{1/n}f_n) \right| = 0 \qquad (10.3.29)$$

uniformly on $\partial \Delta_j$ for all j.

Our next tool will be a certain sequence of approximating polynomials $(q_n)_{n \geqslant 1}$. For each $n \in \mathbb{N}$, choose open sub-arcs γ_j^n of γ_j with one end point at λ_j and such that

$$\lim_{n \to \infty} \omega^* \left(\cup_j \gamma_j^n \right) = 0 \qquad (10.3.30)$$

(see Figure 10.6).

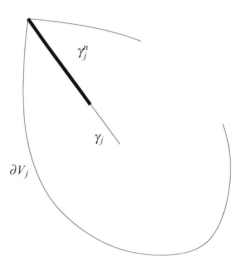

Figure 10.6: The sub-arc γ_j^n and the set $\cup_j (\partial V_j \cup \gamma_j) \setminus \Delta_j$.

Let s_n be a continuous function on

$$\bigcup_j (\partial V_j \cup \gamma_j) \setminus \Delta_j$$

satisfying the three conditions

$$|s_n| \leqslant 1, \quad s_n = \sigma^{1/n} \text{ on } \bigcup_j (\partial V_j \setminus \Delta_j), \quad s_n = 1 \text{ on } \bigcup_j (\gamma_j \setminus \gamma_j^n).$$

Since the domain of definition of s_n has connected complement, we can apply Lavrientiev's theorem [22, p. 232] to produce an analytic polynomial q_n satisfying

$$|s_n - q_n| \leqslant \frac{1}{n(1 + \|f_n\|_{H^2(G)})}.$$

From here we get

$$\int_{\cup_j \partial V_j \setminus \Delta_j} |(\sigma^{1/n} - q_n) Q f_n h_1| ds$$

$$\leqslant \int_{\cup_j \partial V_j \setminus \Delta_j} |(\sigma^{1/n} - s_n) Q f_n h_1| ds + \int_{\cup_j \partial V_j \setminus \Delta_j} |(s_n - q_n) Q f_n h_1| ds$$

$$\leqslant 0 + \frac{1}{n(1 + \|f_n\|_{H^2(G)})} c \|f_n\|_{H^2(G)} \quad \text{(by (10.3.15))}$$

$$\leqslant \frac{c}{n}.$$

Moreover,

$$\int_{\cup_j \gamma_j \setminus \gamma_j^n} |(\sigma^{1/n} - q_n) Q f_n \overline{h}| d\omega^*$$

$$\leqslant \int_{\cup_j \gamma_j \setminus \gamma_j^n} |\sigma^{1/n} - 1| |Q f_n \overline{h}| d\omega^* + \int_{\cup_j \gamma_j \setminus \gamma_j^n} |s_n - q_n| |Q f_n \overline{h}| d\omega^*.$$

By (10.3.3) the first summand is bounded above by

$$c \left(\int_{\cup_j \gamma_j \setminus \gamma_j^n} |\sigma^{1/n} - 1|^2 |h|^2 d\omega^* \right)^{1/2} \left(\int_{\cup_j \gamma_j \setminus \gamma_j^n} |Q f_n|^2 d\omega^* \right)^{1/2}$$

$$\leqslant c \left(\int_{\cup_j \gamma_j \setminus \gamma_j^n} |\sigma^{1/n} - 1|^2 |h|^2 d\omega^* \right)^{1/2} \|f_n\|_{H^2(V)} \quad \text{(by (10.3.7))}$$

which converges to zero as $n \to \infty$ since $|\sigma^{1/n} - 1| \leqslant 2$, $\sigma^{1/n} \to 1$ almost everywhere as $n \to \infty$, and $f_n|V \to g|V$ in $H^2(V)$ norm. Again by (10.3.3) the second summand is bounded above by

$$\frac{1}{n(1 + \|f_n\|_{H^2(G)})} c \|f_n\|_{H^2(G)}$$

which clearly converges to zero as $n \to \infty$. Use the facts that

$$\sup_n \sup \left\{ |q_n(\lambda)| : \lambda \in \bigcup_j (\partial V_j \cup \gamma_j) \setminus \Delta_j \right\} < \infty \quad \text{and} \quad \sup_n \|f_n\|_{H^2(V)} < \infty,$$

along with (10.3.3) and (10.3.7) to get

$$\int_{\cup_j \gamma_j^n} |\sigma^{1/n} - q_n| |Q f_n \overline{h}| d\omega^* \leqslant c \int_{\cup_j \gamma_j^n} |h|^2 d\omega^*$$

which, by elementary measure theory and (10.3.30), goes to zero as $n \to \infty$. Thus we have produced a sequence of analytic polynomials $(q_n)_{n \geqslant 1}$ such that

$$\lim_{n \to \infty} \left(\int_{\cup_j \partial V_j \setminus \Delta_j} |(\sigma^{1/n} - q_n) Q f_n h_1| ds + \int_{\cup_j \gamma_j} |(\sigma^{1/n} - q_n) Q f_n \overline{h}| d\omega^* \right) = 0. \quad (10.3.31)$$

Combine (10.3.31) with (10.3.18) to see that

$$\text{dist}(\lambda, \partial V_j \cap \partial \Delta_j) \left| \frac{1}{2\pi i} \oint_{\cup_j \partial V_j \setminus \Delta_j} \frac{(\sigma^{1/n} - q_n) Q f_n h_1}{z - \lambda} dz + \int_{\cup_j \gamma_j} \frac{(\sigma^{1/n} - q_n) Q f_n \overline{h}}{z - \lambda} d\omega^* \right|$$
$$(10.3.32)$$

goes to zero as $n \to \infty$ uniformly on $\partial \Delta_j$ for all j. Since $\mathcal{M}$ is an invariant subspace, $q_n f_n \in \mathcal{M}$ and so by (10.3.17) $\ell(\cdot, q_n f_n) \equiv 0$ which means that

$$\ell(\cdot, \sigma^{1/n} f_n) = \ell(\cdot, (\sigma^{1/n} - q_n) f_n).$$

For $\lambda \in G \setminus V^-$ we can use (10.3.24) (applied to $g := (\sigma^{1/n} - q_n) f_n$ which belongs to P_V by (10.3.27) and (10.3.20)) to get

$$\ell(\lambda, (\sigma^{1/n} - q_n) f_n) = \frac{1}{2\pi i} \oint_{\cup_j \partial V_j} \frac{Q(\sigma^{1/n} - q_n) f_n h_1}{z - \lambda} dz + \int_{\cup_j \gamma_j} \frac{Q(\sigma^{1/n} - q_n) f_n \overline{h}}{z - \lambda} d\omega^*$$

and so, via (10.3.32),

$$\text{dist}(\lambda, \partial V_j \cap \partial \Delta_j) \left| \ell(\lambda, \sigma^{1/n} f_n) - \frac{1}{2\pi i} \oint_{(\cup_j \partial V_j) \cap (\cup_j \Delta_j)} \frac{Q(\sigma^{1/n} - q_n) f_n h_1}{z - \lambda} dz \right| \quad (10.3.33)$$

goes to zero as $n \to \infty$ uniformly on $\partial \Delta_j \cap (G \setminus V^-)$ for all j. Now use (10.3.28), (10.3.29), and (10.3.33) to get that

$$\text{dist}(\lambda, \partial V_j \cap \partial \Delta_j) \left| \ell(\lambda, g) - \frac{1}{2\pi i} \oint_{(\cup_j \partial V_j) \cap (\cup_j \Delta_j)} \frac{Q(\sigma^{1/n} - q_n) f_n h_1}{z - \lambda} dz \right|$$

goes to zero as $n \to \infty$ uniformly on $\partial \Delta_j \cap (G \setminus V^-)$ for all j. In a similar way,

$$\text{dist}(\lambda, \partial V_j \cap \partial \Delta_j)$$
$$\times \left| \ell(\lambda, g) + Q(\lambda) g(\lambda) h_1(\lambda) - \frac{1}{2\pi i} \oint_{(\cup_j \partial V_j) \cap (\cup_j \Delta_j)} \frac{Q(\sigma^{1/n} - q_n) f_n h_1}{z - \lambda} dz \right|$$

goes to zero as $n \to \infty$ uniformly on $\partial \Delta_j \cap V^-$ for all j.

Let

$$u_n(\lambda) := \frac{1}{2\pi i} \oint_{(\cup_j \partial V_j) \cap (\cup_j \Delta_j)} \frac{Q(\sigma^{1/n} - q_n) f_n h_1}{z - \lambda} dz$$

and notice that

$$u_n \in H^p(\widehat{\mathbb{C}} \setminus \cup_j \Delta_j^-)$$

for all $0 < p < 1$.[1] If F is the $\widehat{\mathbb{C}} \setminus \cup_j \Delta_j^-$-outer function with

$$|F| = \text{dist}(\lambda, \partial V_j \cap \partial \Delta_j) \text{ on } \partial \Delta_j,$$

then our work in the previous paragraph shows that the functions $(Fu_n)_{n \geqslant 1}$ form a Cauchy sequence in $H^2(\widehat{\mathbb{C}} \setminus \cup_j \Delta_j^-)$. then

$$Fu_n \to F\ell(\cdot, g) \text{ almost everywhere on } (\cup_j \partial \Delta_j) \cap (G \setminus V^-);$$

$$Fu_n \to F\ell(\cdot, g) + FQgh_1 \text{ almost everywhere on } (\cup_j \partial \Delta_j) \cap V.$$

By Hardy space theory,
$$F\ell(\cdot, g)|(\cup_j \partial \Delta_j) \cap (G \setminus V^-)$$

is part of the boundary function for a function in $H^2(\widehat{\mathbb{C}} \setminus \cup_j \Delta_j^-)$ and so it follows that $\ell(\cdot, g)|G$ has an analytic continuation to $\widehat{\mathbb{C}} \setminus \cup_j \Delta_j^-$. But, by adjusting the disks Δ_j, we can show that $\ell(\cdot, g)|G$ has an analytic continuation to $\widehat{\mathbb{C}}$ (the Riemann sphere!) and thus must be the constant function.

From (10.3.16) we see that $\ell(\cdot, g)|G$ also has a pseudocontinuation $\ell(\cdot, g)|\mathbb{D}_e$. But by uniqueness of pseudocontinuations (Privalov's uniqueness theorem), the pseudocontinuation and the analytic continuation must be the same. Hence $\ell(\cdot, g)$, originally defined on $\mathbb{D}_e \cup G$, has an analytic continuation to $\widehat{\mathbb{C}}$ which is constant. However, $\ell(\infty, g) = 0$ and so $\ell(\cdot, g) \equiv 0$. In particular, $\ell(\cdot, g) \equiv 0$ on $\mathbb{D}_e$ which implies (10.3.9). $\qquad\square$

10.4   Finally the proof

With the technicalities out of the way, we are now ready for the proof of the main result of this chapter.

Proof of Theorem 10.1.2. Without loss of generality, we assume that the greatest common G-inner divisor of $\mathcal{M}$ is one. Recall that V_j are the open sets guaranteed by Lemma 10.2.1 and

$$V = \bigcup_{j=1}^{n} (V_i \cap G).$$

For $f \in \mathcal{M}$, note that

$$\|f|V\|_{H^2(V)}^2 = \sum_{j=1}^{N} \|f|V_j \cap G\|_{H^2(V_j \cap G)}^2$$

---

[1] Thus far, we have been talking about Hardy spaces of simply connected domains in $\widehat{\mathbb{C}}$. The domain $\widehat{\mathbb{C}} \setminus \cup_j \Delta_j^-$ is not simply connected. However, the known and expected results for the Hardy spaces of simply connected domains still hold for these types of Hardy spaces. We refer the reader to [31, 47, 64, 75] for a discussion of this.

and so if

$$\mathcal{M}_V := \mathrm{clos}_{H^2(V)} \mathcal{M}|V,$$

then

$$\mathcal{M}_j := (\mathcal{M}_V)|(V_j \cap G)$$

is a closed invariant subspace of $H^2(V_j \cap G)$ for each $j = 1, \ldots, N$. Since each V_j is a Carathéodory domain, $\beta_j^{-1} : \mathbb{D} \to V_j$ is a weak-$*$ generator for $H^\infty(\mathbb{D})$ [66]. This means that the polynomials are weak-$*$ sequentially dense in $H^\infty(V_j)$. It follows that $\beta_j \cdot \mathcal{M}_j \subset \mathcal{M}_j$ and so $\mathcal{M}_j \circ \beta_j^{-1}$ is an invariant subspace of $H^2(\mathbb{D} \setminus [0,1))$. (Recall that $\beta_j(\gamma_j \setminus \{\lambda_j\}) = [0,1)$.) Applying Theorem 6.2.1 we obtain $E_j \subset \gamma_j$, $\rho_j : E_j \to \mathbb{C}$, and a $V_j \cap G$-outer function F_j such that

$$\mathcal{M}_j = \left\{ g \in H^2(V_j \cap G) : \frac{g}{F_j} \mid V_j \cap G \in H^2(V_j \cap G), \ g^+ = \rho_j g^- \ \text{a.e. on } E_j \right\}.$$

Now let

$$E := \bigcup_{j=1}^{N} E_j,$$

$$\rho : E \to \mathbb{C}, \quad \rho|E_j := \rho_j,$$

$$F : V \to \mathbb{C}, \quad F|V_j \cap G := F_j.$$

The result now follows from Proposition 10.3.1. $\square$

Chapter 11

Final thoughts

Hardy-Smirnov class: It turns out that our results about the invariant subspaces of $H^2(G)$ can be used to prove analogous results for the Hardy-Smirnov class $E^2(G)$ (recall the definition from (2.1.9)). We restrict our discussion to the case of the slit disk $G = \mathbb{D} \setminus [0,1)$.

If ϕ_G is the conformal map from $\mathbb{D}$ onto G and $\psi = \phi_G^{-1}$, it follows from (2.1.10) that the operator

$$U : H^2(G) \to E^2(G), \quad Uf = (\psi')^{1/2} f$$

is unitary and moreover, U intertwines S ($Sf = zf$) on $E^2(G)$ with S on $H^2(G)$. Letting $K = (\psi')^{1/2}$, we see from the form of ψ' (which can be computed from the appendix) that K is a G-outer function. Furthermore, $KH^2(G) = E^2(G)$.

Thus if $\mathcal{M}$ is an invariant subspace of $E^2(G)$, then $\frac{1}{K}\mathcal{M}$ is an invariant subspace of $H^2(G)$ and so from Theorem 6.2.1, there is an $E \subset [0,1]$, a $\rho : E \to \mathbb{C}$, and, for every $\varepsilon > 0$, a G_ε-outer function F_ε such that

$$\mathcal{M} = K\Theta_{\mathcal{M}} \cdot \left\{ f \in H^2(G) : \frac{f}{F_\varepsilon} \in H^2(G_\varepsilon), f^+ = \rho f^- \text{ a.e. on } E \right\}.$$

Setting

$$\tilde{\rho} := \frac{K^+}{K^-}\rho \quad \text{and} \quad \tilde{F}_\varepsilon = KF_\varepsilon,$$

we see that $\tilde{F}_\varepsilon$ is G_ε-outer and

$$\mathcal{M} = \Theta_{\mathcal{M}} \cdot \left\{ g \in E^2(G) : \frac{g}{\tilde{F}_\varepsilon} \in H^2(G_\varepsilon), g^+ = \tilde{\rho} g^- \text{ a.e. on } E \right\}. \qquad (11.0.1)$$

Banach space case: The main techniques used to characterize the invariant subspaces of $H^2(G)$ depend on our discussion of the nearly invariant subspaces of $H^2(\widehat{\mathbb{C}} \setminus \gamma)$. It is here where we used, in key places such as Proposition 3.4.2 and Proposition 3.4.8, Hilbert space techniques such as the Wold decomposition. These do not have Banach space analogs and so our techniques do not seem to work for $H^p(\widehat{\mathbb{C}} \setminus \gamma)$ for general $p \in [1, \infty)$.

The main results of this monograph (Theorem 3.1.2 and Theorem 6.2.1) should be true in the H^p setting but will require Banach space techniques.

Geometry: Our results characterize the invariant subspaces of $H^2(G)$ for a slit domain where the analytic slit meets the boundary of ∂G at a positive angle. Does anything change when the slit is tangent to ∂G? Can we relax the condition that the slit is analytic? Does the slit domain need to be simply connected? What happens when there are a countably infinite number of slits?

Cyclic invariant subspaces: For certain special $f \in H^2(G)$ (Theorem 7.2.2 and Corollary 7.2.6), we can compute $[f]$, the invariant subspace generated by f. What is $[f]$ for a general $f \in H^2(G)$? We also know (Remark 7.2.1) that not every invariant subspace is cyclic. Which ones are? For example, from Corollary 8.2.11 we see that if $\mathcal{M}$ is cyclic then $\sigma_e(S|\mathcal{M}) = \mathbb{T}$. Is the converse true?

Lattice operations: For any bounded domain Ω, the invariant subspaces of $H^2(\Omega)$ form a lattice in that if $\mathcal{M}_1$ and $\mathcal{M}_2$ are invariant subspaces, then so are $\mathcal{M}_1 \vee \mathcal{M}_2$ and $\mathcal{M}_1 \cap \mathcal{M}_2$. When $\Omega = \mathbb{D}$, we know from Beurling's theorem that $\mathcal{M}_1 = \Theta_1 H^2(\mathbb{D})$ and $\mathcal{M}_2 = \Theta_2 H^2(\mathbb{D})$ for some $\mathbb{D}$-inner functions Θ_1 and Θ_2. Moreover [22, p. 137],

$$\mathcal{M}_1 \vee \mathcal{M}_2 = \text{g.c.d.}(\Theta_1, \Theta_2) H^2(\mathbb{D}), \quad \mathcal{M}_1 \cap \mathcal{M}_2 = \text{l.c.m.}(\Theta_1, \Theta_2) H^2(\mathbb{D}).$$

When $\Omega = G \setminus [0, 1)$, Theorem 6.2.1 says that the invariant subspaces $\mathcal{M}_1$ and $\mathcal{M}_2$ depend on the parameters $\Theta_j, \rho_j, E_j, F_j$, $j = 1, 2$. Can one describe $\mathcal{M}_1 \cap \mathcal{M}_2$ and $\mathcal{M}_1 \vee \mathcal{M}_2$ in terms of these parameters?

Appendix

In this appendix we store some information about two particular conformal maps used in this monograph. We first mention a few words about the conformal map

$$\phi_G : \mathbb{D} \to \mathbb{D} \setminus [0,1).$$

Consider the following sequence of conformal maps

$$w_1(z) := i\frac{1+z}{1-z} : \mathbb{D} \to \{\Im z > 0\};$$

$$w_2(z) := \sqrt{z} : \{\Im z > 0\} \to \{\Im z > 0\} \cap \{\Re z > 0\};$$

$$w_3(z) := \frac{z-1}{z+1} : \{\Im z > 0\} \cap \{\Re z > 0\} \to \mathbb{D} \cap \{\Im z > 0\}.$$

Notice that

$$w_1(\{e^{it} : \pi < t < 3\pi/2\}) = (0,1), \qquad w_1(\{e^{it} : 3\pi/2 < t < 2\pi\}) = (1,\infty);$$
$$w_2(0,1) = (0,1), \qquad w_2(1,\infty) = (1,\infty);$$
$$w_3(0,1) = (-1,0), \qquad w_3(1,\infty) = (0,1).$$

Define

$$\phi_G(z) = (w_3(w_2(w_1(-iz))))^2$$

and notice that ϕ_G is a conformal map from $\mathbb{D}$ onto the slit disk $G = \mathbb{D} \setminus [0,1)$ and, by following the boundaries, we see that ϕ_G maps the arc $\{e^{it} : 0 < t < \pi/2\}$ to the top half of the slit while ϕ_G maps $\{e^{-it} : 0 < t < \pi/2\}$ to the bottom half of the slit. Furthermore, it is routine to show that ϕ_G maps the interval $(-1,1)$ onto $(-1,0)$ and so, by the Schwarz reflection principle, ϕ_G has the following nice property,

$$\phi_G(e^{i\theta}) = \overline{\phi_G(e^{-i\theta})}, \quad 0 \leqslant \theta \leqslant 2\pi.$$

The previous identity says that if $\psi_G = \phi_G^{-1}$, then

$$\psi_G^+(x) = \overline{\psi_G^-(x)}, \quad 0 < x < 1,$$

and so

$$|(\psi_G')^+(x)| = |(\psi_G')^-(x)|, \quad 0 < x < 1.$$

Next, we say a few words about the conformal map

$$\alpha : \mathbb{D} \setminus [0,1) \to \widehat{\mathbb{C}} \setminus \gamma, \quad \gamma = \{e^{it} : -\pi/2 \leqslant t \leqslant \pi\}.$$

We construct this map with a sequence of standard conformal maps:

$$u_1(z) := \frac{1+z}{1-z} : \mathbb{D} \setminus [0,1) \to \{\Re z > 0\} \setminus [1,\infty);$$

$$u_2(z) := z^2 : \{\Re z > 0\} \setminus [1,\infty) \to \widehat{\mathbb{C}} \setminus ((-\infty,0] \cup [1,\infty));$$

$$u_3(z) := \frac{z-i}{z+i} : \widehat{\mathbb{C}} \setminus ((-\infty,0] \cup [1,\infty)) \to \widehat{\mathbb{C}} \setminus \gamma,$$

Observe that

$$u_1(0,1) = (1,\infty),$$

$$u_1(\mathbb{D}_+) = \{\Re z > 0\} \cap \{\Im z > 0\}, \quad u_1(\mathbb{D}_-) = \{\Re z > 0\} \cap \{\Im z < 0\};$$

$$u_2(\{\Re z > 0\} \cap \{\Im z > 0\}) = \{\Im z > 0\}, \quad u_2(\{\Re z > 0\} \cap \{\Im z < 0\}) = \{\Im z < 0\};$$

$$u_3(-\infty,0] = \gamma', \quad \gamma' := \{e^{it} : 0 \leqslant t \leqslant \pi\},$$

$$u_3[1,\infty) = \gamma'', \quad \gamma'' := \{e^{it} : -\pi/2 \leqslant t \leqslant 0\};$$

$$u_3(\{\Im z > 0\}) = \mathbb{D}, \quad u_3(\{\Im z < 0\}) = \widehat{\mathbb{C}} \setminus \mathbb{D}^-.$$

From here, we define

$$\alpha := u_3 \circ u_2 \circ u_1$$

and note that

$$\alpha : \mathbb{D} \setminus [0,1) \to \widehat{\mathbb{C}} \setminus \gamma,$$

$$\alpha[0,1] = \gamma'',$$

$$\alpha(\mathbb{T} \cap \{\Im z > 0\}) = \gamma', \quad \alpha(\mathbb{T} \cap \{\Im z < 0\}) = \gamma',$$

$$\alpha(\mathbb{D}_+) = \mathbb{D}, \quad \alpha(\mathbb{D}_-) = \widehat{\mathbb{C}} \setminus \mathbb{D}^-.$$

Furthermore, it is easy to check from the identity

$$\alpha(z) = \frac{(1+z)^2(1-z)^{-2} - i}{(1+z)^2(1-z)^{-2} + i}$$

that

$$\alpha(e^{-i\theta}) = \alpha(e^{i\theta}).$$

Bibliography

[1] L. Ahlfors and A. Beurling, *Conformal invariants and function-theoretic null-sets*, Acta Math. **83** (1950), 101–129.

[2] J. Akeroyd, *A note concerning cyclic vectors in Hardy and Bergman spaces*, Function spaces (Edwardsville, IL, 1990), Lecture Notes in Pure and Appl. Math., vol. 136, Dekker, New York, 1992, pp. 1–8.

[3] ———, *An extension of Szego's theorem*, Indiana Univ. Math. J. **43** (1994), no. 4, 1339–1347.

[4] J. Akeroyd, D. Khavinson, and H. S. Shapiro, *Remarks concerning cyclic vectors in Hardy and Bergman spaces*, Michigan Math. J. **38** (1991), no. 2, 191–205.

[5] A. Aleman and B. Korenblum, *Derivation-invariant subspaces of C^∞*, Comput. Methods Funct. Theory **8** (2008), no. 1–2, 493–512.

[6] A. Aleman and R. Olin, *Hardy spaces of crescent domains*, preprint.

[7] A. Aleman and S. Richter, *Simply invariant subspaces of H^2 of some multiply connected regions*, Integral Equations and Operator Theory **24** (1996), no. 2, 127–155.

[8] A. Aleman, S. Richter, and C. Sundberg, *Beurling's theorem for the Bergman space*, Acta Math. **177** (1996), no. 2, 275–310.

[9] S. Axler, *Multiplication operators on Bergman spaces*, J. Reine Angew. Math. **336** (1982), 26–44.

[10] S. Axler, P. Bourdon, and W. Ramey, *Harmonic function theory*, Graduate Texts in Mathematics, vol. 137, Springer-Verlag, New York, 1992.

[11] S. Axler, J. B. Conway, and G. McDonald, *Toeplitz operators on Bergman spaces*, Canad. J. Math. **34** (1982), no. 2, 466–483.

[12] F. Bagemihl and W. Seidel, *Some boundary properties of analytic functions*, Math. Z. **61** (1954), 186–199.

[13] A. Beurling, *On two problems concerning linear transformations in Hilbert space*, Acta Math. **81** (1948), 239–255.

[14] L. Brown and A. Shields, *Cyclic vectors in the Dirichlet space*, Trans. Amer. Math. Soc. **285** (1984), no. 1, 269–303.

[15] J. A. Cima, A. L. Matheson, and W. T. Ross, *The Cauchy transform*, Mathematical Surveys and Monographs, vol. 125, American Mathematical Society, Providence, RI, 2006.

[16] J. A. Cima and W. T. Ross, *The backward shift on the Hardy space*, Mathematical Surveys and Monographs, vol. 79, American Mathematical Society, Providence, RI, 2000.

[17] E. F. Collingwood and A. J. Lohwater, *The theory of cluster sets*, Cambridge University Press, Cambridge, 1966.

[18] J. B. Conway, *Functions of one complex variable*, second ed., Graduate Texts in Mathematics, vol. 11, Springer-Verlag, New York, 1978.

[19] _____, *A course in functional analysis*, Graduate Texts in Mathematics, vol. 96, Springer-Verlag, New York, 1985.

[20] _____, *Spectral properties of certain operators on Hardy spaces of planar regions*, Integral Equations and Operator Theory **10** (1987), no. 5, 659–706, Toeplitz lectures 1987 (Tel-Aviv, 1987).

[21] _____, *A course in functional analysis*, second ed., Graduate Texts in Mathematics, vol. 96, Springer-Verlag, New York, 1990.

[22] _____, *The theory of subnormal operators*, Mathematical Surveys and Monographs, vol. 36, American Mathematical Society, Providence, RI, 1991.

[23] _____, *Functions of one complex variable. II*, Graduate Texts in Mathematics, vol. 159, Springer-Verlag, New York, 1995.

[24] _____, *A course in operator theory*, Graduate Studies in Mathematics, vol. 21, American Mathematical Society, Providence, RI, 2000.

[25] L. de Branges, *Hilbert spaces of entire functions*, Prentice-Hall Inc., Englewood Cliffs, N.J., 1968.

[26] R. G. Douglas, H. S. Shapiro, and A. L. Shields, *Cyclic vectors and invariant subspaces for the backward shift operator.*, Ann. Inst. Fourier (Grenoble) **20** (1970), no. fasc. 1, 37–76.

[27] P. Duren, D. Khavinson, and H. S. Shapiro, *Extremal functions in invariant subspaces of Bergman spaces*, Illinois J. Math. **40** (1996), no. 2, 202–210.

[28] P. Duren, D. Khavinson, H. S. Shapiro, and C. Sundberg, *Contractive zero-divisors in Bergman spaces*, Pacific J. Math. **157** (1993), no. 1, 37–56.

[29] _____, *Invariant subspaces in Bergman spaces and the biharmonic equation*, Michigan Math. J. **41** (1994), no. 2, 247–259.

[30] P. Duren and A. Schuster, *Bergman spaces*, Mathematical Surveys and Monographs, vol. 100, American Mathematical Society, Providence, RI, 2004.

[31] P. L. Duren, *Theory of H^p spaces*, Academic Press, New York, 1970.

[32] S. Fisher, *Function theory on planar domains*, Pure and Applied Mathematics (New York), John Wiley & Sons Inc., New York, 1983, A second course in complex analysis, A Wiley-Interscience Publication.

[33] J. B. Garnett, *Analytic capacity and measure*, Springer-Verlag, Berlin, 1972, Lecture Notes in Mathematics, Vol. 297.

[34] _____, *Bounded analytic functions*, Pure and Applied Mathematics, vol. 96, Academic Press Inc. [Harcourt Brace Jovanovich Publishers], New York, 1981.

[35] J. B. Garnett and D. Marshall, *Harmonic measure*, New Mathematical Monographs, vol. 2, Cambridge University Press, Cambridge, 2005.

[36] A. Hartmann, D. Sarason, and K. Seip, *Surjective Toeplitz operators*, Acta Sci. Math. (Szeged) **70** (2004), no. 3–4, 609–621.

[37] M. Hasumi, *Invariant subspace theorems for finite Riemann surfaces*, Canad. J. Math. **18** (1966), 240–255.

[38] _____, *Hardy classes on infinitely connected Riemann surfaces*, Lecture Notes in Mathematics, vol. 1027, Springer-Verlag, Berlin, 1983.

[39] E. Hayashi, *Classification of nearly invariant subspaces of the backward shift*, Proc. Amer. Math. Soc. **110** (1990), no. 2, 441–448.

[40] W. K. Hayman, *On functions with positive real part*, J. London Math. Soc. **36** (1961), 35–48.

[41] H. Hedenmalm, *A factorization theorem for square area-integrable analytic functions*, J. Reine Angew. Math. **422** (1991), 45–68.

[42] H. Hedenmalm, B. Korenblum, and K. Zhu, *Theory of Bergman spaces*, Graduate Texts in Mathematics, vol. 199, Springer-Verlag, New York, 2000.

[43] H. Helson, *Lectures on invariant subspaces*, Academic Press, New York, 1964.

[44] D. Hitt, *Invariant subspaces of H^2 of an annulus*, Pacific J. Math. **134** (1988), no. 1, 101–120.

[45] K. Hoffman, *Banach spaces of analytic functions*, Dover Publications Inc., New York, 1988, Reprint of the 1962 original.

[46] M. V. Keldysh and M. A. Lavrientiev, *Sur la représentation conforme des domaines limités par des courbes rectifiables*, Ann. Sci. École Norm. Sup. **54** (1937), 1–38.

[47] D. Khavinson, *Factorization theorems for different classes of analytic functions in multiply connected domains*, Pacific J. Math. **108** (1983), no. 2, 295–318.

[48] D. Khavinson and H. S. Shapiro, *Invariant subspaces in Bergman spaces and Hedenmalm's boundary value problem*, Ark. Mat. **32** (1994), no. 2, 309–321.

[49] S. Ja. Khavinson, *Removable singularities of analytic functions of the V. I. Smirnov class*, Some problems in modern function theory (Proc. Conf. Modern Problems of Geometric Theory of Functions, Inst. Math., Acad. Sci. USSR, Novosibirsk, 1976) (Russian), Akad. Nauk SSSR Sibirsk. Otdel. Inst. Mat., Novosibirsk, 1976, pp. 160–166.

[50] A. V. Kiselev and S. N. Naboko, *Nonself-adjoint operators with almost Hermitian spectrum: matrix model. I*, J. Comput. Appl. Math. **194** (2006), no. 1, 115–130.

[51] P. Koosis, *Introduction to H_p spaces*, second ed., Cambridge Tracts in Mathematics, vol. 115, Cambridge University Press, Cambridge, 1998, With two appendices by V. P. Havin [Viktor Petrovich Khavin].

[52] B. I. Korenblum, *Invariant subspaces of the shift operator in a weighted Hilbert space*, Mat. Sb. (N.S.) **89 (131)** (1972), 110–137, 166.

[53] _____, *Invariant subspaces of the shift operator in certain weighted Hilbert spaces of sequences*, Dokl. Akad. Nauk SSSR **202** (1972), 1258–1260.

[54] N. Makarov and A. Poltoratski, *Meromorphic inner functions, Toeplitz kernels and the uncertainty principle*, Perspectives in analysis, Math. Phys. Stud., vol. 27, Springer, Berlin, 2005, pp. 185–252.

[55] A. I. Markushevich, *Theory of functions of a complex variable. Vol. I, II, III*, english ed., Chelsea Publishing Co., New York, 1977, Translated and edited by Richard A. Silverman.

[56] Ch. Pommerenke, *Boundary behaviour of conformal maps*, Grundlehren der Mathematischen Wissenschaften, vol. 299, Springer-Verlag, Berlin, 1992.

[57] I. I. Privalov, *Randeigenschaften analytischer Funktionen*, Zweite, unter Redaktion von A. I. Markuschewitsch überarbeitete und ergänzte Auflage. Hochschulbücher für Mathematik, Bd. 25, VEB Deutscher Verlag der Wissenschaften, Berlin, 1956.

[58] T. Ransford, *Potential theory in the complex plane*, London Mathematical Society Student Texts, vol. 28, Cambridge University Press, Cambridge, 1995.

[59] S. Richter, *Invariant subspaces in Banach spaces of analytic functions*, Trans. Amer. Math. Soc. **304** (1987), no. 2, 585–616.

[60] _____, *Invariant subspaces of the Dirichlet shift*, J. Reine Angew. Math. **386** (1988), 205–220.

[61] S. Richter and C. Sundberg, *Multipliers and invariant subspaces in the Dirichlet space*, J. Operator Theory **28** (1992), no. 1, 167–186.

[62] _____, *Invariant subspaces of the Dirichlet shift and pseudocontinuations*, Trans. Amer. Math. Soc. **341** (1994), no. 2, 863–879.

[63] W. T. Ross and H. S. Shapiro, *Generalized analytic continuation*, University Lecture Series, vol. 25, American Mathematical Society, Providence, RI, 2002.

[64] H. L. Royden, *Invariant subspaces of H^p for multiply connected regions*, Pacific J. Math. **134** (1988), no. 1, 151–172.

[65] D. Sarason, *The H^p spaces of an annulus*, Mem. Amer. Math. Soc. No. **56** (1965), 78.

[66] _____, *Weak-star generators of H^∞*, Pacific J. Math. **17** (1966), 519–528.

[67] _____, *Nearly invariant subspaces of the backward shift*, Contributions to operator theory and its applications (Mesa, AZ, 1987), Oper. Theory Adv. Appl., vol. 35, Birkhäuser, Basel, 1988, pp. 481–493.

[68] S. Shimorin, *Wold-type decompositions and wandering subspaces for operators close to isometries*, J. Reine Angew. Math. **531** (2001), 147–189.

[69] N. A. Shirokov, *Analytic functions smooth up to the boundary*, Lecture Notes in Mathematics, vol. 1312, Springer-Verlag, Berlin, 1988.

[70] B. Simon, *Spectral analysis of rank one perturbations and applications*, Mathematical quantum theory. II. Schrödinger operators (Vancouver, BC, 1993), CRM Proc. Lecture Notes, vol. 8, Amer. Math. Soc., Providence, RI, 1995, pp. 109–149.

[71] V. I. Smirnov, *Sur les formules de Cauchy et Green et quelques problèms qui s'y rattachent*, Izv. Akad. Nauk SSSR Ser. Mat. **VII** (1932), no. 3, 331–372.

[72] V. I. Smirnov and N. A. Lebedev, *Functions of a complex variable: Constructive theory*, Translated from the Russian by Scripta Technica Ltd, The M.I.T. Press, Cambridge, Mass., 1968.

[73] C. Sundberg, *Analytic continuability of Bergman inner functions*, Michigan Math. J. **44** (1997), no. 2, 399–407.

[74] G. C. Tumarkin and S. Ja. Havinson, *On the removing of singularities for analytic functions of a certain class (class D)*, Uspehi Mat. Nauk (N.S.) **12** (1957), no. 4 (76), 193–199.

[75] G. C. Tumarkin and S. Ja. Khavinson, *Classes of analytic functions in multiply connected regions represented by Cauchy-Green formulas*, Uspehi Mat. Nauk (N.S.) **13** (1958), no. 2 (80), 215–221.

[76] M. Voichick, *Ideals and invariant subspaces of analytic functions*, Trans. Amer. Math. Soc. **111** (1964), 493–512.

[77] _____, *Invariant subspaces on Riemann surfaces*, Canad. J. Math. **18** (1966), 399–403.

[78] D. V. Yakubovich, *Invariant subspaces of the operator of multiplication by z in the space E^p in a multiply connected domain*, Zap. Nauchn. Sem. Leningrad. Otdel. Mat. Inst. Steklov. (LOMI) **178** (1989), no. Issled. Linein. Oper. Teorii Funktsii. 18, 166–183, 186–187.

Index